Concepts and Applications of Oilseeds

Concepts and Applications of Oilseeds

Edited by **Shirley Doy**

New York

Published by Callisto Reference,
106 Park Avenue, Suite 200,
New York, NY 10016, USA
www.callistoreference.com

Concepts and Applications of Oilseeds
Edited by Shirley Doy

© 2015 Callisto Reference

International Standard Book Number: 978-1-63239-119-3 (Hardback)

Contents

Preface

This book has been an outcome of determined endeavour from a group of educationists in the field. The primary objective was to involve a broad spectrum of professionals from diverse cultural background involved in the field for developing new researches. The book not only targets students but also scholars pursuing higher research for further enhancement of the theoretical and practical applications of the subject.

The various concepts and applications of oilseeds are elucidated in this profound book. It has been compiled with the objective of providing in-depth information on the extensive topic of oilseeds. It consists of descriptive accounts on oil presses and oilseed pests with comprehensive information about sesame seeds. The book discusses a wide spectrum of topics related to oilseeds including a discussion regarding nitrogen efficiency in oilseed rape with emphasis on its physiological mechanism. The book has been compiled with information contributed by researchers and authors from across the globe. The ultimate objective of this book is to serve as a valuable source of reference for researchers, students, and general readers interested in attaining extensive knowledge about oilseeds.

It was an honour to edit such a profound book and also a challenging task to compile and examine all the relevant data for accuracy and originality. I wish to acknowledge the efforts of the contributors for submitting such brilliant and diverse chapters in the field and for endlessly working for the completion of the book. Last, but not the least; I thank my family for being a constant source of support in all my research endeavours.

Editor

Prospects for Transgenic and Molecular Breeding for Cold Tolerance in Canola (*Brassica napus* L.)

Anthony O. Ananga[1], Ernst Cebert[2], Joel W. Ochieng[3],
Suresh Kumar[2], Devaiah Kambiranda[4], Hemanth Vasanthaiah[4],
Violetka Tsolova[1], Zachary Senwo[2], Koffi Konan[5] and Felicia N. Anike[6]

1. Introduction

Oilseed rape has become a major crop in North America, with cropland dedicated to rapeseed production increasing from 4,391,660 ha in 2001 to 7,103,725 ha in 2010 in both U.S.A. and Canada (Canola Connection, 2011; National Agricultural Statistics Service, 2011). Most of these are cultivated in spring in the Canadian Prairie Provinces and the northern Great Plains of the USA.

Canola is cultivated both during winter and spring seasons in the United States and this exposes the crop to winter kill, frost, and high temperatures, during the reproductive period. The temperatures during winter and spring are known to influence all the crucial steps of the reproductive cycle including gametogenesis, pollination, fertilization and embryogenesis (Angadi, 2000). Winter rapeseed has been successfully grown in the Pacific Northwest, southern Great Plains, Midwest, and southeast regions of the USA. The hardiest cultivars will routinely survive winters in the north east of USA but survival is inconsistent further south (Rife *et al.*, 2001). Winter-grown canola (*Brassica napus* L.) production is limited mostly by frost and winter-kill in the southern canola-growing regions of the United States (Singh *et al.*, 2008). For instance, the late freeze in 2007 resulted in significant damage to most of the winter canola cultivars at the National Winter Canola Variety Trials in Alabama, U.S. (Cebert and Rufina, 2007). Winter hardiness and freezing tolerance are a major concern for improving production consistency in many regions of the canola growing countries.

[1]CESTA, Center for Viticulture and Small Fruit Research, Florida A&M University, Tallahassee, USA
[2]Department of Biological and Environmental Sciences, Alabama A&M University, Normal, USA
[3]Faculties of Agriculture and Veterinary Medicine, University of Nairobi, Nairobi, Kenya
[4]Plant Biotechnology Laboratory, College of Agriculture, Florida A & M University, Tallahassee, USA
[5]Department of Food and Animal Sciences, Food Biotechnology Laboratory,
Alabama A&M University, Normal, USA
[6]Department of Natural Resources and Environmental Design, North Carolina A&T State University
Greensboro, USA

Introduction and cultivation of new crops in a given environment require management practices and trait selection that enable optimum performance of the crop. Canola is an important oilseed crop and its cultivation is expanding, particularly in the western world because of its importance as both an oilseed and a bio-diesel crop.

1.1 Cold tolerance

The ability of rapeseed plants to tolerate very low temperatures depends essentially on their development and the degree of hardening it has achieved. Unhardened plants can survive -4°C, while fully-hardened spring-type rapeseed can survive much lower temperatures (-10° to -12°C). Hardened winter rapeseed can survive short periods of exposure to temperatures between 15° and -20°C. Unhardening happens fairly fast after the plants initiate active growth (Sovero, 1993). The plants are typically best adapted to survive the winter in rosette stage with 6 to 8 leaves. Small plants are usually not as capable of surviving over-wintering, while plants with more leaves often start the stem elongation prematurely, exposing the meristem tissue to cold, making it more susceptible to damage. The hardening requirements of rapeseed have not been fully characterized. Winter type canola tend to harden faster, achieve higher degree of cold tolerance and unharden slower than spring types, but it is likely that variable hardening requirements could also be found within both types. Some differences in cold hardiness have been observed among both winter and spring types, however, it is unclear whether these are due to differences in ultimate achievable cold hardiness or differences in hardening requirements. The absence of snow cover during the coldest period of the winter decreases the plants' chances to survive. Ice formation on the soil surface can damage the crown area of the plants and reduce survival rate (Sovero, 1993).

1.2 Vernalization requirement

Most winter rapeseed cultivars will require three weeks of near-freezing temperatures in the field to get fully vernalized and start rapid generative growth. In controlled environments, eight weeks at 4°C temperature is sufficient for full vernalization. In spring planting, winter rape will typically start slow generative growth after the prolonged rosette stage, and some cultivars may start blooming towards the end of the growing season. Differences in this respect are sometimes useful in distinguishing between morphologically similar cultivars. Different vernalization requirements are apparent among winter rape cultivars. A high vernalization requirement does not necessarily result in good winter hardiness, as many of the winter type cultivars from extreme maritime environments, such as Japan, require a long vernalization period yet have little tolerance for low temperatures (Sovero, 1993).

Some spring type cultivars do not exhibit any vernalization response at all, but in some cases the generative development can be accelerated with brief chilling treatment. In spring planting, only a few cool nights are usually needed. Vernalization response in spring types also tends to disappear in a long day environment. In spite of the variability in vernalization requirements within both types, the differences between the two types i.e. winter and spring canola are fairly clear with no overlap in the initiation of blooming in either spring or fall planting (Sovero, 1993).

1.3 Cold stress symptoms and its after effects

Cold stress symptoms can arise only after a cold temperature event; however, mild symptoms of herbicide injury may often be confused with symptoms caused by cold stress temperatures or nutrient-deficient soil (Figure 1 to 3). Recovery from cold stress will be rapid as temperatures increase. Nutrient stress symptoms are unlikely to occur at the cotyledon stage as nutrient demands at this stage are generally low (Boyles. 2011).

Fig. 1 shows that since the 1st and 2nd leaves are of normal size, the purpling observed is not herbicide injury. The purpling is as a result of anthocynin production caused by cold temperatures. Purpling may be towards the base, on the leaf margins or may cover entire young leaves of the plant. This symptom will diminish as temperatures increase.

Fig. 2 exhibits cupping caused by cold temperatures and symptoms quickly diminish as temperatures increases.

Fig. 3 indicates that cupping was caused by a low level herbicide residue. Variation in herbicide carryover means uninjured (red arrow) and injured yellow arrow and plants may be found in close proximity. Cold stress generally causes more uniform damage.

Fig. 1. Purpling of leaves due to low temperature.

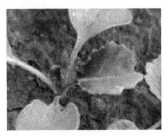

Fig. 2. Cupping of leaves due to low temperature.

Fig. 3. Leaves damaged by herbicide carryover.

1.4 Methods of measuring cold stress

Field sites often exhibit either complete survival or complete winter-kill. Because of this variability, laboratory procedures to measure freezing tolerance have been developed including plant tissue water content (Brule-Babel and Fowler, 1988), ion leakage from plant cells after a freezing stress (Teutonico *et al.*, 1993), and changes in luminescence (Brzostowicz and Barcikowska, 1987). Further, meristem regrowth after plants are subjected to freezing temperatures is commonly used (Andrews and Morrison, 1992).

Laboratory freezing tolerance procedures have allowed investigators to gather information on cold tolerance that would otherwise be unobtainable in the field. Freezing tolerance of plant tissue is evaluated by measuring whether the tissues are alive or dead after subjecting the tissue to a range of freezing temperatures. The extent of damage caused by the freezing can be evaluated by placing plant tissue in distilled water and measuring the electrical conductivity of the resultant solution (Madakadze *et al.*, 2003; Murray *et al.*, 1989). An increased rate of electrolyte loss is interpreted as evidence and the extent corresponds to damage. The electrolyte leakage (EL) method (Oakton CON 510TDS electrical conductivity meter: Eutech Instruments, Singapore) is based on objective measurements, which utilizes small quantities of tissue, and is relatively cheap. However, it takes more time than the chlorophyll fluorescence method (OS1-FL portable pulse-modulated fluorometer: Opti-Sciences, Tynggsboro, MA).

Freezing on the functionality of the photosynthetic apparatus can be used to assess the cold tolerance of plant genotypes (Chengci *et al.*, 2005). The photosynthetic apparatus function can be evaluated by measuring the ratio of chlorophyll variable fluorescence (Fv) over the maximum fluorescence value (Fm) (F_v/F_m), which indicates the efficiency of the excitation capture by open photosystem II reaction centers (Frachebound *et al.*, 1999; Rizza *et al.*, 2001). A significant reversible decrease in F_v/F_m was found in all genotypes of oat (*Avena sativa* L.) during acclimation to low, nonfreezing temperatures, and F_v/F_m measurement was found to be highly correlated with field-evaluated frost damage (Rizza *et al.*, 2001). Measurement of F_v/F_m is rapid and noninvasive.

2. Effect of cold stress on plant performance

2.1 Cold stresses reduce plant productivity

To make early spring seeding feasible, suitable canola cultivars must be selected. The suitable cultivars must have quick germination emergence, and establishment at low temperatures, and seedlings must be tolerant to early spring freezing and thawing events. Low temperatures reduce both the final percentage as well as the rate of germination, which leads to delayed and reduced seedling emergence of canola (Zheng et al., 1994). Early spring frosts are more problematic with fall-seeded canola, which emerges earlier than canola seeded in early spring (Willenborg *et al.*, 2004). Fall seeding of canola does present some challenges in the Canadian Prairie Provinces and the northern Great Plains of the USA. Fall seeding frequently entails seeding into hard, cold soil, which ultimately results in poor soil-to-seed contact (Kirkland and Johnson, 2000).

Mild winter climate in Northern Alabama, USA is conducive for optimum productivity of winter canola (Cebert and Rufina, 2007). However, the region is also vulnerable to late frost during mid to late spring (late March to Mid April). In April, 2007, twenty winter canola genotypes at varying stages of flowering and early pod-filling (Figures 4-9) were exposed to three incidences of naturally occurring severe spring frosts that reduced most of the yield components (Cebert and Rufina, 2007).

Fig. 4-9 show the extent of damage resulting from cold stress after freeze from April 5 to 9, 2007 in North Alabama, U.S.A.

Fig. 4. Canola field at flowering after spring 2007 hard freeze in Northern Alabama.

Fig. 5. Early maturity breeding line (lower right) completely destroyed by late spring frost.

Fig. 6. Several cultivars did not suffer much damage. Cultivar, Kadore produced the highest yield despite the freeze followed by an exceptional drought period.

Fig. 7. Primary yield loss due to dropping off of fertilized flowers after late spring frost.

Fig. 8. Destruction of photosynthetic green tissues due to late spring frost.

Fig. 9. Destruction of developing pods due to late spring frost.

Early seeded canola may encounter suboptimal soil temperatures for seed germination and seedling establishment in the northern Great Plains (Chengci *et al.*, 2005). The optimum germination temperature for canola was reported to be 5°C (Morrison *et al.*, 1989) and 0.4 to 1.2°C (Vigil *et al.*, 1997). Soil temperature in April in most canola production areas of the northern Great Plains is usually lower than the previously given optimum temperatures (Zheng *et al.*, 1994). Spring canola seeded into suboptimal soil temperatures had lower emergence and stand establishment rates due to the seed rotting in the cold soils (Blackshaw, 1991; Livingston and de Jong, 1990). A significant reduction in canola germination was found at temperatures less than 10°C (Nykiforuk and Johnson-Flanagan, 1994), and it took as long as 18 days for 50% emergence at 5°C (Blackshaw, 1991). Vigil *et al.* (1997) reported that 65 to 81 growing degree days (GDD) are required for spring canola seedlings to emerge. Differences in GDD have also been found among the species and seed lots (Nykiforuk and Johnson-Flanagan, 1994)

One can determine an optimal seeding date in the northern Great Plains or other regions with similar climate and soil conditions as central Montana (Chengci *et al.*, 2005). First, days from seeding to emergence can be predicted from long-term weather data based on the base temperature for germination (T_b) and growing degree days for 50% emergence. Second, based on the cold tolerance information, one can decide which cultivar to plant and the risk of frost damage for a given early seeding date. Third, using the information on days to 50 % flowering in combination with long-term weather data, the maximum daily temperatures at flowering stage, optimum seeding date can be forecasted; thus, the potential impact of maximum temperature on seed yield can be estimated for a given seeding date.

2.2 Plant responses to cold stress

Plants acclimatize to survive metabolic lesions because of intracellular ice formation, as well as to survive the dehydrative effects of frost (Kacperska, 1984). Fowler *et al.*, (1996) found that after the vernalization requirement was met in wheat (*Triticum aestivum* L.) and rye (*Secale cereale* L.), cold acclimation declined. Laroche *et al.*, (1992) did not observe this reduction in cold acclimation in rapeseed but they estimated cell survival on excised leaves to determine freezing tolerance and not crown meristem survival.

The relationship between vernalization requirements and cold tolerance is not clear as different observations have been reported. While Markowski and Rapacz (1994) found little relationship between these traits by comparing vernalization requirements and frost resistance of winter rape lines derived from doubled haploid, Rapacz and Markowski (1999) found a significant correlation between vernalization requirement and both frost resistance and field survival when looking at older, high erucic acid cultivars. Long vernalization requirements are expected to delay a plant from entering the reproductive growth phase, a cold sensitive plant growth stage (Fowler et al., 1996). Rife and Zeinali (2003) found that rapeseed plants may withstand cold temperatures under field conditions more effectively prior to vernalization saturation than after the vernalization requirement has been met.

Under field conditions, rapeseed plants often survive cold events in December and January only to be killed by less severe cold events in February and March (Rife and Zeinali, 2003). One theory to explain this is that after vernalization saturation takes place, rapeseed plants do not have the same ability to recover after a warming event as unvernalized plants. This has been documented in winter cereals. Fowler et al., (1996) found reductions in lethal temperature 50's (LT$_{50}$s) of 5°C or more for many cultivars between Day 49 and 84 of acclimation at 4°C. This study suggested that this may not be the case in three rapeseed cultivars: A112, Ceres, and Plainsman. However, the spike of increased cold tolerance was more pronounced in unvernalized seedlings. Under the variable temperature conditions present during Great Plains winters, this phenomenon could have a substantial impact on winter survival before vernalization saturation and selecting genotypes with increased vernalization requirements could have a positive effect on winter survival.

Early seeding between late March and mid April has been proposed as the key to achieve a good and stable canola seed yield in central Montana, USA (Chengci et al., 2005). This study indicated that the optimal seeding rate for early spring seeded canola is 32 to 65 seeds m^{-2}. Although early spring seeded canola is expected to encounter cold soil temperatures and frequent frosts, canola can germinate at less than 4°C and requires 42 to 81 growing degree days (GDD) for 50% flowering for emergence. Further studies are needed to test the threshold temperature and duration of canola genotypes to cold stress. Several genotypes were found to have favorable characteristics for the semi-arid region in the northern Great Plains, such as low T_b, fast emergence, relatively cold tolerant, early flowering, and good seed yield and oil content. Further, Chengci et al., (2005) observed that a high F_v/F_m and low EL reading after a snowstorm event indicated less effect on Photosynthetic System II and cell membranes by cold stress. A large recovery of F_v/F_m and slow leakage rate (longer time to reach 50% total leakage) also indicate a frost tolerance of the plants. There were variances observed among the cultivars in F_v/F_m, $\Delta(F_v/F_m)$, EL, and hours to 50% total leakage (HTL$_{50}$). However, as the authors noted, there was no evidence of cell membrane damage by the snowstorm, EL readings ranged from 12.7 to 20.4 µS cm^{-1}, and the time to reach 50% of the total leakage ($\approx$600 µS cm^{-1} after autoclaving) ranged from 212 to 276 h. Several cultivars had greater F_v/F_m readings and less $\Delta(F_v/F_m)$ than others. Neither EL nor the F_v/F_m readings were correlated with seed yield. However, the positive correlations between $\Delta(F_v/F_m)$ and biomass indicated that cultivars having faster biomass growth may be sensitive to cold stress. Results in this study also indicated that canola did not suffer severe frost damage by the snowstorm on 12 May 2004. Canola seedlings at the early seeding dates encountered several snowstorms in May and early June over 3 yr from 2002 to 2004, but no severe frost damage to the seedlings was observed. Canola was also found to have the ability to withstand subzero temperatures in other studies after reclamation. Kirkland and Johnson (2000) found fall-seeded canola survived eight consecutive nights of frost with the temperature dropping as low as −8°C in 1994. Further, Johnson et al. (1995) observed that canola seedlings were able to tolerate temperatures of −6°C without significant reductions in plant stands in North Dakota

2.3 Effect of cold stress during flowering and pod formation

Farmers in central Alberta experienced a record string of killing frosts for their Canola plants in late May to early June, 2000. Seedling canola was severely injured by these frosts and significant reseeding occurred. The surviving density was on average 32 to 43 plants/m^2 and about 11/m^2 in the worst areas. Continuing frosts hampered the recovery of the canola seedling but eventually new growth appeared after 10 days. The canola stand had a slow recovery for a few weeks following damage, but initially looked quite poor. By flowering, the stand began to fill in but differences in maturity were evident between areas that suffered different amounts of stand thinning. The stand continued to improve through flowering and pod fill. The crop matured in September with the thin areas maturing later than the better areas by one to two weeks. Plant counts at harvest averaged 43 plants/m². The thin crop that was questioned in spring eventually yielded 2,128 kg/ha gross with 2 to 3% green. Although this crop probably would have yielded higher if frost had not occurred, the yield was satisfactory and equivalent or better than reseeded canola. Early spring germinated stands can suffer damage due to subsequent heavy spring frosts. Polymer-coated seeds show promise to reduce the untimely germination but carry extra costs (Canola Council of Canada, 2011). However, Christian *et al.*, (2004)) observed that by decreasing osmotic potential and temperature, germination significantly reduced on both coated and uncoated canola seeds. Polymer-coated seeds exhibited delayed germination even in the absence of moisture stress, an effect that was magnified at more negative water potentials and at a lower temperature. Median germination time of polymer-coated seed was significantly higher than for uncoated control seed throughout all temperature and osmotic potential treatments.

Several types of polymer seed coats have recently been developed with the intent to extend the fall planting period (Zaychuk and Enders, 2001). The polymers are specifically designed to prevent germination of canola seeds until spring by absorbing water into the polymer coat matrix but preventing the passage of sufficient amounts of water to the seed coat to begin germination. After water entry into the polymer coat matrix, freezing is required to create microfractures in the polymer coat that act as water channels for imbibition. Polymer seed coats have been shown to decrease imbibition (Chachalis and Smith, 2001) and final germination percentages (Valdes and Bradford, 1987). This may also be a problem with the polymer coats developed specifically for fall seeding canola as Gan et al., (2001) observed reduced canola emergence during a dry spring following a dry fall. Lower yields were also observed in polymer coat treatments.

To make early spring seeding feasible, suitable canola cultivars must be selected. The suitable cultivars must have quick germination, emergence, and establishment at low temperatures, and seedlings must be tolerant to early spring freezing and thawing events. Freezing on the functionality of the photosynthetic apparatus can be used to assess the cold tolerance of plant genotypes. The photosynthetic apparatus function can be evaluated by measuring the F_v/F_m, which indicates the efficiency of the excitation capture by open Photosystem II reaction centers (Rizza et al., 2001). Rizza *et al* (2001) observed a significant reversible decrease in F_v/F_m in all genotypes of oat (*Avena sativa* L.) during acclimation to low, nonfreezing temperatures, and F_v/F_m measurement was found to be highly correlated with field-evaluated frost damage.

The late freeze of April 5-9, 2007, in Northern Alabama, resulted in significant economic damage to most of the winter canola cultivars that were evaluated for the National Winter Canola Variety Trials at the Hazel Green, Alabama location (Cebert and Rufina, 2007). Five consecutive nights of low temperatures: 0.6, 1.1, -5.6, -5.6, and 0.6°C with chilling hours of 8, 16, 18, 24, and 17occurred from April 5-9, 2007, respectively. These dates correspond to days 192, 193, 194, 195, and 196 after planting, when all cultivars were beyond 50% blooming (Figures 4-9). Damage on the primary stem was highest in early blooming cultivars (Figure 5). Cultivars which were in full bloom between days 178-183 (March 22-27) suffered a complete loss. Seeds per pod and days to 50% blooming were the two factors other than the extent of freeze damage which influenced seed yield (Table 1). In Alabama

Cultivar	Estimated freeze damage (%)	Estimated yield (Kg ha-1)
Baros	80.0a†	374.6i
Virginia	65.0ab	685.7hi
Viking	60.0abc	838.9fghi
Hybrista	56.7abc	782.8ghi
Abilene	55.0abcd	851fghi
Taurus	53.3abcd	902.7fghi
Trabant	53.3abcd	893.6fghi
Baldur	40.0bcde	1195.6defgh
Rasmus	35.0bcdef	972.2fgh
Falstaff	28.3cdef	1332.5defg
Summer	26.7cdef	924.7fghi
Ceres	21.7def	1399.2cdef
Wichita	21.7def	1721.2bcd
Satori	16.7ef	1392.3cdef
Jetton	16.7ef	1061.8efgh
Kalif	16.7ef	2094.6ab
Ovation	10.0ef	1578bcde
Kronos	8.3ef	1624.8bcd
Kadore	1.7f	2552.0a
Plainsman	1.7f	1900.6bc

†Means with the same letter are not significantly different at a = 0.05

Table 1. Estimated freeze damage and yield (Kg ha-1) of canola cultivars following late spring frost damage and exceptional drought conditions in National Winter Canola Variety Trial, Alabama A&M University, USA 2007.

A&M University, U.S. experimental plots, Cebert and Rufina (2007) reported an inverse response among cultivars between freeze damage and seed yield. Cultivars Kadore, Kalif and Plainsman with the highest seed yields were also the last ones to reach 50% blooming, approximately five days before the freeze.

3. Breeding for crop improvement

3.1 Genetics

As described by Nakashima and Yamaguchi-Shinozaki (2006), the basic characteristics of plants forced them to survive in environments with variable environmental stresses such as cold stress and osmotic stress, which includes drought and high salinity. Plants, therefore exhibit an increase in freezing tolerance in response to low, non-freezing temperatures. This concept of acclimating to stressful environmental conditions has been widely investigated, especially with the use of model plant *Arabidopsis thaliana* (Yamaguchi-Shinozaki and Shinozaki, 1994; Jaglo-Ottosen *et al.*, 1998; Lee *et al.*, 2001; Taji *et al.*, 2002; Zhang *et al.*, 2004b; Carvallo *et al.*, 2011).

However, contribution towards the improvement of *Brassica* species, which eventually led to the development of canola, is the creation of the Triangle of U (Figure 10) by Nagaharu, U (1935). The interpretation of three common diploid species (*B. campestris, B. nigra* and *B. oleracea*) being the origin of the current tetraploid *Brassica* species continues to be the foundation for contemporary development and enhancement for both vegetables and oilseed crops within the genus, indicative in improvement of *B. rapa* by Ofori *et al.*, (2008). An extensive assortment of loci for various physiological and morphological traits in the different *Brassicica* species have been determined and identified through both classical and molecular methods. Through the use of embryo rescue techniques, interspecific and intergeneric hybridization hindrances such as male-sterility and self-incompatibility were overcome (Nishi *et al.*, 1959; Quazi, 1988; Mohaptra and Bajaj, 1987). As delineated by Harberd (1969), the tedious procedures for embryo rescue in *Brassica* species resulted in the exchange of genes to produce cultivars with improved resistance for pests, variation for seed coat color, changes in fatty-acids composition and tolerance to herbicide. Sacristan and Gerdemann (1986), successfully transferred blackleg resistance to *B. napus* from *B. juncea* using embryo culture, while similar techniques were successfully employed to transfer both aphid-resistance and triazine herbicide-resistance among and between the various *Brassica* species (Kott *et al.*, 1988).

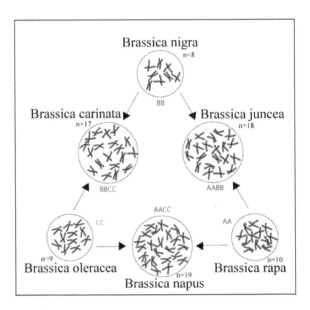

Fig. 10. The Triangle of U, by Woo Jang-choon (Nagaharu U, 1935).

Microspore culture and protoplast fusion are other protocols which have been employed successfully in the development of improved *Brassica* cultivars. According to results obtained by Ryschka *et al.*, (2007), hybrid formation between haploid at the level of protoplasts fusion obtained from different species has the potential to combine divergent genomes which may not be possible otherwise. Traditional non-molecular systems provided the foundation to identify Quantitative Trait Loci (QTLs) for early genomic mapping. These systems of hybridization coupled with current biotechnology have energized the search for novel procedures to create distinctive desirable cultivars.

3.2 Breeding for cold tolerance

Current molecular tools such as SSRs (Single Sequence Repeats), SNPs (Single Nucleotide Polymorphism) and ESTs (Expressed Sequence Tags) are being used to identify genes of economic significance including cold tolerance in *Brassica* species as reported in studies by Ofori *et al.*, (2008) and Thomashow (1999). Specific genes that respond to cold and osmotic stress in plants have been elucidated by Zhang *et al.*, (2004a), and Shinozaki *et al.*, (2003) and summarized by Lata and Prasad (2011). Results from these findings along with molecular makers have enabled the creation of high density genetic and physical maps of new genes that will allow the enhancement of genetic variation for desired traits such as response to cold stress. The most recent of such results is the release of the *Brassica rapa* genome-sequence by the "*Brassica* Genome Sequencing Project Consortium" (2011). Initiatives to identify stress related genes have provided significant results in understanding functional genomics of abiotic stress. The Genome Canada/Genome Prairie project studied a range of genomics and proteomics technologies to determine how plants respond to various environmental stresses at the gene level, particularly to cold. A general insight among the findings is the discernment that cold tolerance genes are induced under conditions other than low temperatures but also due to dehydration, high salt and other abiotic stress.

Many studies (Sangwan *et al.*, 2001; Zhu *et al.*, 2009; Chinnusamy *et al.*, 2010; Chen *et al.*, 2011) have reported significant findings that provide novel sources of cold tolerance genes which can be introgressed into new hybrids and open pollinated cultivars. New data are being generated with genomics tools to exploit the use of genetic maps derive from various *Brassica* species and those from *Arabidopsis*. The accumulation of ESTs and SNPs is producing significant information on genome polymorphism and sequence data for all stress related characteristics in *Brassica* species. Kreps *et al.*, (2002) used transcriptome changes for *Arabidopsis* to identify genes in response to stress treatment including cold. Their findings showed approximately 30% of the transcriptones demonstrated sensitivity to regulation to common stress, with most being unambiguous in response for specific stimuli. Genes identified as circadian controlled were also found to be associated in response to cold stress stimuli. Similar findings were also reported by Trischuk *et al.*, (2006) and Nakashima and Yamaguchi-Shinozaki (2006) who further indicated that cold acclimation, while primarily influenced by temperature, is also moderated by factors such as light intensity, day length, cultural practices, and other abiotic stresses such as drought, dehydration and salinity. Dehydration responsive element binding (DREBs) have been identified as important plant transcription factors (TFs) in regulating expression of many stress-inducible genes. These DREBs elements have been identified in many plant species as indicated in Table 2 in their response to cold stress (Lata & Prasad, 2011).

DREB TFs	Species	Accession no.	Stress response	References
DREB1A	*Arabidopsis thaliana*	AB007787	Cold	Liu *et al.*, 1998
DREB2C	*Arabidopsis thaliana*	At2g40340	Salt, Mannitol, Cold	Lee *et al.*, 2010
CBF1	*Arabidopsis thaliana*	U77378	Cold	Gilmour *et al.*, 1998
CBF2	*Arabidopsis thaliana*	AF074601	Cold	Gilmour *et al.*, 1998
CBF3	*Arabidopsis thaliana*	AF074602	Cold	Gilmour *et al.*, 1998
OsDREB1A	*Oryza sativa*	AF300970	Cold, Salt, Wounding	Dubouzet *et al.*, 2003
OsDREB1B	*Oryza sativa*	AF300972	Cold	Dubouzet *et al.*, 2003
OsDREB1C	*Oryza sativa*	AP001168	Drought, Salt, Cold, ABA, Wound	Dubouzet *et al.*, 2003
OsDREB2A	*Oryza sativa*	AF300971	Drought, Salt, faintly to Cold, ABA	Dubouzet *et al.*, 2003
OsDREB1F	*Oryza sativa*		Drought, Salt, Cold, ABA	Wang *et al.*, 2008
OsDREB2B	*Oryza sativa*		Heat, Cold	Matsukura *et al.*,2010
OsDREBL	*Oryza sativa*	AF494422	Cold	Chen *et al.*, 2003
TaDREB1	*Triticum aestivum*	AAL01124	Cold, Dehydration, ABA	Shen *et al.*, 2003
WCBF2	*Triticum aestivum*		Cold, Drought	Kume *et al.*, 2005
WDREB2	*Triticum aestivum*	BAD97369	Drought, Salt, Cold, ABA	Egawa *et al.*, 2006
HvDREB1	*Hordeum vulgare*	DQ012941	Drought, Salt, Cold	Xu *et al.*, 2009
ZmDREB2A	*Zea mays*	AB218832	Drought, Salt, Cold, Heat	Qin *et al.*, 2007
PgDREB2A	*Pennisetum glaucum*	AAV90624	Drought, Salt, Cold	Agarwal *et al.*, 2007
GmDREBa	*Glycine max*	AY542886	Cold, Drought, Salt	Li *et al.*, 2005
GmDREBb	*Glycine max*	AY296651	Cold, Drought, Salt	Li *et al.*, 2005
PpDBF1	*Physcomitrella patens*	ABA43697	Drought, Salt, Cold, ABA	Liu *et al.*, 2007
PNDREB1	*Arachis hypogea*	FM955398	Drought, Cold	Mei *et al.*, 2009
DvDREB2A	*Dendrathema*	EF633987	Drought, Heat, ABA, Cold	Liu *et al.*, 2008
DmDREBa	*Dendronthema3moriforlium*	EF490996	Cold, ABA	Yang *et al.*, 2009
DmDREBb	*Dendronthema3moriforlium*	EF487535	Cold, ABA	Yang *et al.*, 2009
PeDREB2	*Populus euphratica*	EF137176	Drought, Salt, Cold	Chen *et al.*, 2009

*Adapted from: Lata and Prasad (2011).

Table 2. Transcription factors identified as regulators in the expression abiotic stress including cold tolerance*

3.3 Limitations of classical breeding

Brassica species having characteristics such as sporophytic pollen self-incompatibility, male sterility along with restoration for cytoplasmic male sterility offered overwhelming possibilities for genetic modification. The production of haploids and doubled haploids using microspores has accelerated the production of homozygous lines in the *Brassica* species (Cardoza and Stewart, 2004). Somatic cell fusion has facilitated the development of interspecific and intergeneric hybrids in the sexually incompatible species of *Brassica*. Those characteristics have been further exploited by combining traditional non-molecular protocols with modern biotechnology including molecular markers in marker-assisted selection and breeding, and transformation technology to introduce desired genes into elite cultivars.

However, classical breeding which relies largely on homologous recombination between chromosomes to generate genetic diversity (Garcia *et al.*, 1998), limits the extent of further improvement as it does not allow the precise understanding of genomic composition. The continuation of current technology in identifying selectable markers to aid in the assortment of segregating population is a major contribution that will extend the use of classical breeding. However, the use of plant transformation which allows the isolation and insertion of single genes into elite cultivars could accelerate crop improvement once the technology becomes routine.

4. Molecular breeding for crop improvement

4.1 Genomic approach

Functional genomics strategies to understand a plant's response to abiotic stress and to exploit this knowledge to improve crop yield, quality & quantity under adverse environmental conditions is a priority focus for northern temperate climates. Genomic approaches are likely to have particular value for *Brassica* crop improvement because they have the potential to identify transcriptional, biochemical, and genetic pathways that contribute to agronomic properties. Examples include revealing transcriptional pathways that are correlated with oil quality and disease resistance (e.g. specific resistance genes and downstream transcriptional pathways). The application of such knowledge to *Brassica* crop improvement program is likely to take the form of improved cultural practices and precise molecular breeding. Approaches such as marker-assisted selection and transgenesis will facilitate transfer of genes for desirable traits into elite or classic cultivars of *Brassica*, with the goal of improving agronomic performance while preserving traditional quality traits.

A complete genome sequence provides unlimited information in the sequenced organism as well as in the related taxa (Yang *et al.*, 2005). Korea *Brassica* Genome Project (KBGP) in conjunction with Multinational *Brassica* Genome Project (MBGP) have sequenced chromosome 1 (cytogenetically oriented chromosome #1) of *Brassica rapa*. They selected 48 seed BACs on chromosome 1 using EST genetic markers and FISH analyses. They also reported that the comparative genome analyses of the EST sequences and sequenced BAC clones from *Brassica* chromosome 1 revealed homeologous partner regions on the *Arabidopsis* genome and a syntenic comparative map between *Brassica* chromosome 1 and *Arabidopsis* chromosomes. In-depth sequence analyses of five homeologous BAC clones and an *Arabidopsis* chromosomal region revealed overall co-linearity, with 82% sequence similarity.

The data indicated that the *Brassica* genome has undergone triplication and subsequent gene losses after the divergence of *Arabidopsis* and *Brassica*.

Cheung *et al.* (2009) analyzed homoeologous regions of *Brassica* genomes at the sequence level. These represented segments of the *Brassica* A genome as found in *Brassica rapa* and *Brassica napus* and the corresponding segments of the *Brassica* C genome as found in *Brassica oleracea* and *B. napus*. Analysis of synonymous base substitution rates within modeled genes revealed a relatively broad range of times (0.12 to 1.37 million years ago) since the divergence of orthologous genome segments as represented in *B. napus* and the diploid species. Similar and consistent ranges were also identified for single nucleotide polymorphism and insertion-deletion variation. Genes conserved across the *Brassica* genomes and the homoeologous segments of the genome of *Arabidopsis thaliana* showed almost perfect collinearity. Numerous examples of apparent transduplication of gene fragments, as previously reported in *B. oleracea*, were also observed in *B. rapa* and *B. napus*, indicating that this phenomenon is widespread in *Brassica* species. They also concluded that the majority of the regions studied, the C genome segments were expanded in size relative to their A genome counterparts. Further, they observed considerable variation, even between the different versions of the same *Brassica* genome, for gene fragments and annotated putative genes suggesting that the concept of the pan-genome might be particularly appropriate when considering *Brassica* genomes. Thus characterization of complete *Brassica* genome and comparative genome analyses with *Arabidopsis* using genomics approach will increase the knowledge of the biological mechanisms of the *Brassica* species that will allow targeted approaches to reduce the number and impact, which could enable a sustainable and environmentally-sound, farming policy.

4.1.1 Molecular markers

Molecular markers have been used extensively to analyze genetic diversity and to create linkage maps in *Brassica* crops. Most importantly, they have been used widely to map agronomically important genes in *Brassica* genomes and to assist canola breeding and selection procedures. According to Snowdon and Friedt (2004), the major challenges that face *Brassica* geneticists and breeders are (i) the alignment of existing genetic and physical maps in order to agree on a consensus map with a standardized nomenclature, and (ii) to compile and integrate relevant phenological, morphological, and agronomic information with allelic information for *Brassica* oilseed germplasm. Therefore, association mapping can be used to exploit available genetic resources outside the narrow rapeseed gene pool. Molecular markers linked to agronomically important traits have been reported, and most of them are now integrated into oilseed breeding programmes. It is important to note that genome research and marker assisted applications in *Brassica* began to flourish in the late 1980s (Snowdon and Freidt, 2004), when the first restriction fragment length polymorphism (RFLP) linkage maps for *B. oleracea* (Slocum *et al.*, 1990), *B. rapa* (Song *et al.*, 1991), and *B. napus* (Landry *et al.*, 1991) were developed. The development of PCR techniques over the last two decades has lead to the rise of new marker technologies, and this has enabled the generation of high-density molecular maps through the amplification of highly polymorphic anonymous PCR fragments. Some of the DNA-based marker systems that have been used in *Brassica* are random amplified polymorphic DNA markers (RAPD; Williams et al., 1990), amplified fragment length polymorphisms (AFLP; Vos *et al.*, 1995), and inter-simple sequence repeats (ISSR; Zietkiewicz *et al.*, 1994).

PCR-based markers that meet the requirements and capacity of rapeseed breeders has emerged due to the ability to convert anonymous PCR markers that are closely linked to loci controlling traits of interest into sequence characterized amplified region (SCAR) or sequence tagged site (STS). Progress has also been made with the development of simple sequence repeat markers, also termed microsatellites (SSR; Grist *et al.*, 1993). These markers are highly polymorphic, robust, and relatively inexpensive. They are valuable to use for map alignment among different crosses simply because of their co-dominant nature. The number of publicly available *Brassica* SSR primers is increasing as a result of publicly funded international initiatives (www.brassica.info/ssr/SSRinfo.htm). Another important marker system is the single-nucleotide polymorphisms (SNPs), which results from single-base substitutions in the DNA sequence. These are the most abundant form of DNA polymorphism in most organisms, and they offer an opportunity to develop extremely fine genetic maps. This is because they can be used to uncover allelic variation directly within the expressed sequences, and to develop haplotypes based on gametic phase disequilibrium for analysis of quantitative traits.

The application of all these marker techniques towards developing maps for cold tolerance in *Brassica* crops will greatly improve the production of canola in temperate regions. Collaborative research with many research groups to improve stress tolerance in canola by utilizing these marker systems should be encouraged through funding. The results obtained through these collaborative studies can contribute to the sustainable oil and food production in canola.

4.1.2 Gene expression during cold stress in *Brassica* species

Exposure of cold-hardy species to low, non-freezing temperatures induces genetic, morphological and physiological changes in plants, which results in the development of cold hardiness and the acquisition of freezing tolerance (Savitch *et al.*, 2005). The ability of plants to acquire freezing tolerance from cold acclimation has been shown to involve the reprogramming of gene expression networks (Seki *et al.*, 2001, Fowler and Thomashow, 2002; Kreps *et al.*, 2002; Seki *et al.*, 2002). Photosynthetic cold acclimation has been reported to be an essential component of the development of cold hardiness and freezing tolerance and requires the complex interaction of low temperature, light and chloroplast redox poise (Gray *et al.*, 1997, Wanner and Junttila, 1999).

The long-term cold acclimation has also been shown to be associated with morphological changes, such as compact dwarf morphology, increased leaf thickness caused by increased mesophyll cell size, increased specific leaf weight, marked decrease of leaf water content and an increase in cytoplasmic volume relative to vacuole volume (Stefanowska *et al.*, 1999; Strand *et al.*, 1999; Stefanowska *et al.*, 2002). In fact, it has been suggested that such structural changes might be necessary to account for the increase in stromal and cytosolic enzymes and metabolites in cold-acclimated leaves (Strand *et al.* 1999).

Several genes associated with cold hardiness have been studied in various crops. But in *Brassica napus*, limited report exists. Savitch *et al.* (2005) studied the effects of over expression of two *Brassica* CBF/ DREB1-like transcription factors (BNCBF5 and 17) in *Brassica napus* cv. Westar. They reported that in addition to developing constitutive freezing tolerance and constitutively accumulating COR gene mRNAs, BNCBF5 and 17 over expressing plants also accumulate moderate transcript levels of genes involved in photosynthesis and chloroplast development as identified by microarray and Northern analyses. These include GLK1 and GLK2-like transcription factors involved in chloroplast photosynthetic development, chloroplast stroma cyclophilin ROC4 (AtCYP20–3), β-amylase and triose-P/Pi translocator. In parallel with these changes, increases in photosynthetic efficiency and capacity, pigment pool sizes, increased capacities of the Calvin cycle enzymes, and enzymes of starch and sucrose biosynthesis, as well as glycolysis and oxaloacetate/malate exchange were seen, suggesting that BNCBF over expression has partially mimicked cold-induced photosynthetic acclimation constitutively. Taken together, they suggested that BNCBF/DREB1 over expression in *Brassica* not only resulted in increased constitutive freezing tolerance but also partially regulated chloroplast development to increase photochemical efficiency and photosynthetic capacity.

The role of hsp90 in adaptation to cold temperature stress has also been studied. Characterization of the expression of hsp90 genes of *Brassica napus* using northern blot analysis and immunoblotting have shown that the hsp90 mRNA and protein are present in all *B. napus* tissues examined, albeit at different levels (Krishan *et al.*, 1995). High levels of hsp90 mRNA and protein were found in young and rapidly dividing tissues such as shoot apices and flower buds, suggesting that hsp90 may have an important role in plant growth and development. A significant increase in hsp90 mRNA levels was detected in seedlings exposed to 5°C. The transcript levels reached a maximum within 1 d of cold treatment and remained elevated for the entire duration of cold treatment. The levels of hsp90 mRNA rapidly decreased to the level found in control plants upon return to 20°C. The cold-induced accumulation of hsp90 mRNA closely resembles the expression of two previously identified cold-regulated genes of *B. napus*. Further, determining the cellular localization of the above genes and proteins during cold acclimation and identifying the proteins associated with them will provide more clues to the cellular basis of cold tolerance.

4.1.3 Microarray based monitoring of gene expression during cold stress

Genome wide transcription analysis in response to stresses is essential to providing the basis of effective engineering strategies to improve stress tolerance in *Brassica* crop plants (Lee et al., 2008). In order to perform transcriptome analysis in *Brassica rapa*, Lee et al. (2008) constructed a *B. rapa* oligo microarray, KBGP-24K, using sequence information from approximately 24,000 unigenes and analyzed cold (4 degrees C), salt (250 mM NaCl), and drought (air-dry) treated *B. rapa* plants. Among the *B. rapa* unigenes represented on the microarray, 417 (1.7%), 202 (0.8%), and 738 (3.1%) were identified as responsive genes that were differently expressed 5-fold or more at least once during a 48-h treatment with cold, salt, and drought, respectively. These results were confirmed by RT-PCR analysis. In the abiotic stress responsive genes identified, they found 56 transcription factor genes and 60 commonly responsive genes. The authors suggested that various transcriptional regulatory mechanisms and common signaling pathway are working together under the abiotic stresses in *B. rapa*. In conclusion, they reported that their new developed 24K oligo microarray will be a useful tool for transcriptome profiling and this work will provide valuable insight in the response to abiotic stress in *B. rapa* and other *Brassica* species.

4.1.4 Micro RNA could modify regulator gene expression during cold in *Brassica* species

MicroRNAs (miRNAs) are short ribonucleic acid (RNA) molecules (on average only 22 nucleotides long) found in all eukaryotic cells except fungi, algae, and marine plants. miRNAs are post-transcriptional regulators that bind to complementary sequences on target messenger RNA transcripts (mRNAs), usually resulting in translational repression or target degradation and gene silencing. Several miRNAs have been identified that regulate complex process. Kutter et al. (2007) demonstrated that the density and development of stomatal complexes on the epidermis of Arabidopisis thaliana leaves depends on the microRNA-mediated regulation of Agamous-like16 (AGL16), which is a member of the MADS box protein family. AGL16 mRNA is targeted for sequence-specific degradation by miR824, a recently evolved microRNA conserved in the Brassicaceae and encoded at a single genetic locus. They reported that expression of a miR824-resistant AGL16 mRNA, but not the wild type AGL16 mRNA, in transgenic plants increased the incidence of stomata in higher order complexes. By contrast, reduced expression of AGL16 mRNA in the agl16-1 deficiency mutant and in transgenic lines over expressing miR824 decreased the incidence of stomata in higher order complexes. Non overlapping patters of AGL16 mRNA and miR824 localization led to the proposal that the miR824/ AGL16 pathway functions in the satellite meristemoid linage of stomatal development. Since *Brassica* and *Arabidopsis*, derived from same line and diverged around 12 to 20 million years ago, similar conserved miRNA families can be identified in *Brassica* species which help in plant development. Hence, microRNAs so identified will play an essential function in regulating gene expression both in mutlicellular plants and animals.

4.2 Proteomic approach

In the past few decades, considerable efforts have been directed at identifying cold-regulated protein in plant species. Various biochemical responses of plants to low, freezing temperatures have been widely documented. They involve changes in protein content and enzyme activities, metabolic modifications and changes in lipid composition and membrane structure. It is also recognized that *Brassica* species are of value for investigating important key areas for resistance, especially cold resistance, and the great advances have been made in terms of cold-induced genes and cytological mechanisms in the shape of cold resistance (Jiang *et al.*, 1996; Diaz *et al.*, 1997; Mieczyslaw, 1999; Rapacz, 2002). Cold acclimation is a complex adaptive process by which plants increase their tolerance to extracellular freezing. This process is induced by exposure of the plant to low but nonfreezing temperatures and is accompanied by a variety of biochemical and structural changes in plant cells (Thomashow, 2001). Proteins encoded by cold-regulated genes have an interesting feature in common; they are hydrophilic and remain soluble upon boiling (Gilmour *et al.*, 2000). It is postulated that proteins that display these distinctive properties may have roles in cryoprotection (Thomashow, 2001). Other cold-regulated gene products show similarities to antifreeze proteins or to drought-induced late-embryogenesis-abundant proteins (Gilmour *et al.*, 2000) and responsive to ABA proteins which suggests a possible role in preventing ice formation and imparting tolerance to dehydration stress, respectively.

4.2.1 Protein expression during cold stress in *Brassica* species

B. napus, a member of the Cruciferae family, cold acclimates just like *Arabidopsis thaliana*. Jaglo *et al.* (2001) showed that *B. napus* has a cold-response pathway related to the CBF cold-response pathway of *Arabidopsis*. cDNA clones encoding two different CBF-like proteins were identified by screening *B. napus* cDNA libraries using PCR-generated probes. The *B. napus* CBF-like proteins were 92% identical in amino acid sequence to each other and approximately 76% identical in sequence to *Arabidopsis* CBF1. An alignment of the *B. napus* proteins with *Arabidopsis* CBF1 indicated that the sequence identity extended throughout the protein, but was greatest in the AP2/EREBP DNA-binding domain includes an alignment of one *B. napus* CBF protein against *Arabidopsis* CBF1). A sequence for a third *B. napus* CBF polypeptide has been deposited by others (accession no. AF084185; N. Zhou, G. Wu, Y.-P. Gao, R.W. Wilen, and L.V. Gusta). Transcripts encoding *B. napus* CBF-like proteins were found to accumulate rapidly (within 30 min) upon exposure of plants to low temperature. They reasoned that if *B. napus* had a similar CBF-like cold-response pathway, then expression of the *Arabidopsis* CBF genes in transgenic *B. napus* might also activate expression of *Bn115* and other cold-regulated genes containing the CRT/DRE-related regulatory elements and increase plant freezing tolerance. Constitutive expression of *Arabidopsis* CBF1, CBF2, and CBF3 in transgenic *B. napus* caused the accumulation of transcripts for *Bn115* and *Bn28* without a low temperature stimulus; *Bn28* encodes an ortholog of the CRT/DRE-regulated cold-responsive gene *COR6.6* (Hajela *et al.*, 1990). Electrolyte leakage experiments indicated that expression of the *Arabidopsis* CBF genes in *B. napus* resulted in an increase in freezing tolerance. The experiments presented above indicated that *B. napus* encodes a CBF cold-response pathway related to that found in *Arabidopsis*.

There is emerging evidence that certain novel hydrophilic and late embryogenesis abundant (LEA) polypeptides participate in the stabilization of membranes against freeze-induced injury. These hydrophilic and late embryogenesis abundant polypeptides are predicted to contain regions capable of forming amphipathic α-helices which are shown to have strong effect on intrinsic curvature of monolayers and their propensity to form hexagonal II phase. They are said to defer their formation at lower temperatures (Epand *et al.*, 1995). An additional hypothesis suggests that the extensive water binding capacity of these hydrophilic proteins might provide a protective environment in the proximity of membranes during freezing and result in membrane stabilization. In addition, there is evidence that protein denaturation occurs in plants at low temperature (Guy *et al.*, 1998) which could potentially result in cellular damage. In these cases, the enhancement of antioxidative mechanisms (Aroca *et al.*, 2003), increased levels of sugars in the apoplastic space (Livingston and Henson, 1998), and the induction of genes encoding molecular chaperones (Guy *et al.*, 1998), respectively, could have protective effects.

Recently (Chen *et al.*, 2011) studied a gene encoding novel cold-regulated protein with molecular mass of 25 KDa that was isolated from *B. napus* cDNA library using microarray analysis, and is consequently designated as BnCOR25. The data presented in this study revealed that BnCOR25 transcripts were significantly accumulated in roots after cold treatment. Sumoylation/desumoylation of proteins has been shown to have a pivotal role in cold acclimation (Miura *et al.*, 2007). Sumoylation is a post-translational protein modification where small ubiquitin-related modifier (SUMO) proteins are conjugated to protein substrates in a process dependent on SUMO E3 ligases, whereas desumoylation is the removal of SUMO proteins from their target proteins by SUMO proteases. It might protect target proteins from proteasomal degradation because sumoylation prevents ubiquitination (Ulrich, 2005). SIZ1, an *Arabidopsis* SUMO E3 ligase is shown to be required for the accumulation of SUMO conjugates during cold stress.

Transgenic attempts with many structural genes have also been made with a moderate degree of success. The overexpression of genes encoding LEA proteins can improve the stress tolerance of transgenic plants. Expression of the citrus gene encoding a LEA protein, CuCOR19 increased the cold tolerance of transgenic tobacco (Hara *et al.*, 2003). Likewise, the freezing tolerance of *Arabidopsis* was increased by the ectopic expression of the wheat gene WCS19 (Dong *et al.*, 2002), the *Arabidopsis* gene COR15A (Artus *et al.*, 1996), and the co-expression of the genes RAB18 and COR47, and XERO2 and ERD10 (Puhakainen *et al.*, 2004). The freezing tolerance of strawberry leaves was enhanced by expression of the wheat dehydrin gene WCOR410 (Houde *et al.*, 2004). On the other hand, the expression of two cold-induced LEA proteins from spinach (Kaye *et al.*, 1998) and three desiccation-induced LEA proteins from *C. plantagineum* (Iturriaga *et al.*, 1992) in tobacco did not induce any significant changes in the freezing or drought tolerance of the respective transgenic plants. This may indicate either that not all LEA proteins make a significant contribution to plant stress tolerance, or that they need a particular background to function in, as suggested for transgenic strawberry plants (Houde *et al.*, 2004).

5. Using transgenic approaches to develop cold tolerance

Biotechnology offers new strategies that can be used to develop transgenic canola plants with improved tolerance to cold stress. Rapid advancement in recombinant DNA technology and development of precise and efficient gene transfer protocols have resulted in transformation and regeneration of transgenic lines in canola and other plant species (Wani *et al.*, 2008; Gosal *et al.*, 2009; Wani *et al.*, 2011). Genes that respond to freezing stress have been isolated and characterized, and studies suggest that they contribute to chilling tolerance and cold acclimation (Knight *et al.*, 1999; Hsieh *et al.*, 2002). Therefore, the transgenic approach should be pursued actively in canola breeding to improve cold tolerance. Efforts have been made to generate transgenic lines, which have shown improved tolerance to cold stress (Savitch, *et al.*, 2005).

Transgenic technology has the ability to improve cold stress in plants by introducing or down-regulating genes that regulate specific trait (Kumar, 2006). In the last few years, several efforts have been made to identify and characterize cold-responsive (*COR*) genes. Homologous components of the *Arabidopsis CBF* cold response pathway have also been found in many plants (Yamagachi-Shinozaki and Shinozaki, 2006). Some of these putative orthologs have been analyzed, and functionally tested. Most of the studies indicate that the expressions of *CBFs* and *CORs* in response to cold stress are similar in many plant species. Thus, they involve rapid cold induced expression of the *CBFs* followed by expression of *CBF*-targeted genes that increase freezing tolerance. It is important to note that as we aspire to engineer cold stress in Canola, the contitutive expression of *Arabidopsis CBF* genes in other plants resulted in increasing freezing tolerance (Yamagachi-Shinozaki and Shinozaki, 2006). There are also other structural genes that have been used to engineer cold tolerance in some plants with amoderate degree of success. An example is tobacco that was engineered by over-expressing chloroplast glycerol-3-phosphate acyltransferase (*GPAT*) gene from squash and *Arabidopsis* (Murata *et al.*, 1992). The transgenic tobacco showed enhanced cold tolerance and an increase in the number of unsaturated fatty acids present in the plant cell wall. In another study, Pennycooke *et al.*, (2003) down-regulated α-Gal (α –Galactosidase) in petunia, and this resulted in transgenic plants with an increased freezing tolerance. This suggested that transformation with α-Gal is another way in which freezing tolerance of plants can be genetically improved. The genes encoding *LEA* proteins can also improve tolerance to cold stress if they are overexpressed in other plants. Citrus gene encoding a *LEA* protein, *CuCOR19* was over-expressed in tobacco and an increased cold tolerance of transgenic tobacco was achieved (Hara *et al.*, 2003). The expression of the wheat dehydrin gene WCOR410, in strawberry leaves also enhanced freezing tolerance (Houde *et al.*, 2004). In a separate study, Kim *et al.*, (2007) engineered tobacco with ring zinc finger protein (RDCPt) from hot pepper and their results indicated that the expression of this gene improved cold tolerance in transgenic plants when compared to wild type. In another study Su *et al.*, (2010), determined that *MYBS3* was critical in cold adaptation in rice and it enhanced cold tolerance. Their report indicated that transgenic rice constitutively overexpressing *MYBS3* tolerated 4°C for at least 1 week and there were no interferences with the yield. All these studies demonstrate that the relationships among different pathways regulated by cold acclimation are complex. Therefore, it is important to understand the mechanism regulating cold-regulated genes in order to engineer cold tolerant canola. In developing transgenic canola, one can study the genes aforementioned with the aim of over-expressing or down-regulating them in Canola plants. The advent of molecular genetics and biotechnology offers a possibility to genetically engineer Canola to be more tolerant to cold. The technology has been modified to significantly improve breeding efficiency, thus resulting in rapid and accurate incorporation of cold tolerant genes into Canola plants.

5.1 ABA-independent gene regulation to cold stress

Environmental stresses induce the expression of many genes that can be classified into two groups. The first group corresponds to proteins involved in transduction pathways, such as transcription factors, whereas the second group includes effector proteins like the enzymes of osmolyte biosynthesis. Many studies have been focused on transcription factors involved in gene expression regulation. For each signal (salt, drought, and cold), several pathways can be distinguished depending on ABA dependent and ABA independent. ABA-independent expression of stress-responsive genes can occur through dehydration-responsive element (DRE)/C-repeat (CRT) cis-acting elements. The binding factors CBF/DREB1 (CRT-binding factor/DRE-binding factor 1) and DREB2 mediate gene expression in response to cold and drought/salinity, respectively). Interestingly, the CBF4 protein seems to mediate drought response unlike the other CBFs (Haake *et al.*, 2002). A particular feature of CBF proteins is their early and transient cold induction, which precedes the expression of cold-responsive genes. This requires the involvement of a constitutively expressed CBF-transcriptional inducer, which would be activated by cold treatment.

Four orthologues of the *Arabidopsis* CBF/DREB transcriptional activator genes were identified from the winter *Brassica napus*, cv. Jet neuf (Gao *et al.*, 2002). All four BNCBF clones encode a putative DRE/CRT (LTRE)-binding protein with an AP2 DNA-binding domain, a putative nuclear localization signal and a possible acidic activation domain. Deduced amino acid sequences suggested that BNCBFs 5, 7, and 16 are very similar to the *Arabidopsis* CBFI whereas BNCBF17 is different in that it contains two extra regions of 16 and 21 amino acids in the acidic domain. Transcripts hybridizing specifically to BNCBF17 and to one or more of the other BNCBFs accumulated in leaves within 30 min of cold exposure of the *Brassica* seedlings and preceded transcript accumulation of the cold-inducible BN28 gene, a *Brassica* orthologue of the cor6.6 or KIN gene from *Arabidopsis*. Cold-induced accumulation of BNCBF17 mRNA was rapid but was short-lived compared to transcripts hybridizing to BNCBF5/7/16. Transcripts hybridizing to one or more of BNCBF5/7/16 accumulated at low levels after the plants were subjected to prolonged exposure to salt stress. BNCBF17 was not responsive to salt stress. BNCBF transcript accumulation was similar in both spring and winter *Brassica* but the persistence of the transcripts in the cold were generally shorter in the spring than in the winter type. BNCBF5 and 17 proteins bind in vitro to the LTRE domains of the cold-inducible BN115 (cor15a orthologue) or BN28 promoters. Differential binding preferences, however, to LTREs between BNI 15 and BN28 were observed. Mutation of the core CCGAC sequence of the LTRE indicated that BNCBF17 had a lower sequence binding specificity than BNCBF5. Furthermore, experiments indicated that the LTREs were able to drive BNCBF5 and 17 trans-activation of the Lac-Z reporter gene in yeast. We conclude that the BNCBFs reported here could function as trans-acting factors in low-temperature responses in *Brassica*, controlling the expression of cold-induced genes through an ABA-independent pathway. Chen *et al.* (2011) indicated that BnCOR25 protein expression was up-regulated by cold, dehydration, and exogenous ABA treatment, suggesting this gene may be activated via ABA-dependent signal pathway. In this study, the data revealed that BnCOR25 was localized on cell periphery, at which the BnCOR25 protein may bind to cell membrane. In addition, BnCOR25 protein also displays a number of dehyrins' features. Thus, BnCOR25 protein may play a role in cold resistance in a similar manner. It can be concluded, however, that DRE/CRT element, Bn CBFs and Bn COR25 are capable of mediating the cold induced expression of genes in *Brassica napus* via an ABA-independent pathway.

6. Future research

The development of genetically engineered cold tolerant Canola plants by introducing or overexpressing selected genes is a viable option for producing an elite plant. It is also the only option if genes of interest originate from different species or distant relatives. However, the increasing availability of data related to large-scale genome homology between *Arabidopsis* and *Brassica* species means that canola is well-positioned to be among the first major crop species to benefit from continuous progress in plant biotechnology and molecular marker technologies. The sequencing of *B. napus* genome and the emergence of detailed physical maps are of great importance towards engineering cold tolerant canola. The availability of complete gene sequences for *Brassica* can enhance advances in detecting polymorphisms for many agronomically important candidate genes for cold tolerance. It is also possible to elucidate the genetic control of cold tolerance through allele-trait association studies. This can be implemented by combining SNP haplotype data for *Brassica* candidate genes with pedigree and quantitative trait information. DNA-chip and high-throughput SNP genotyping technologies will accelerate our understanding of cold tolerance in canola through molecular genetics.

7. Conclusions

Cold is an environmental factor that adversely affects the productivity and oil quality and limits the geographical distribution and growing season of canola. Although significant progress has been made to elucidate the genetic mechanisms underlying cold tolerance in Canola, our current understanding is limited to single and shallow temperature stress. In cold regions such as Canada, canola plants are subjected to intense levels of cold stresses, and hence, the response of canola to a combination of different cold temperatures deserves more attention. Newly developed varieties offer an opportunity to test the effects of multiple cold temperatures, and to perform extensive field studies under diverse environments to assess their tolerance. Discoveries in cold stress response in model species such as *Arabidopsis thaliana* can be adapted to canola to identify candidate genes, a key element in molecular plant breeding in canola.

8. References

Andrews, J., & Morrison, J. (1992). Freezing and ice tolerance tests for Winter *Brassica*. *Journal of the American Society of Agronomy*, Vol. 84, No. 6, pp. 960-962.

Angadi, V. (2000). Response of three *Brassica* species to high temperature stress during reproductive growth. *Can. J. Plant Sci.*, Vol. 80, pp.693-701.

Aroca, R.; Irigoyen, J. & Sanchezdiaz, M. (2003). Drought enhances maize chilling tolerance.II. Photosynthetic traits and protective mechanisms against oxidative stress. *Physiologia Plantarum*, Vol. 117, pp. 540-549.

Artus, N., Uemura, M., Steponkus, L., Gilmour, J., Lin, C., & Thomashow, F. (1996). Constitutive expression of the cold regulated *Arabidopsis thaliana* COR15a gene affects both chloroplast and protoplast freezing tolerance. *Proceedings of National Academy of Sciences*, Vol. 93, pp. 13404–13409.

Blackshaw, E. (1991). Soil temperature and moisture effects on downy brome vs. winter canola, wheat and rye emergence. *Crop Sci.* Vol. 31, pp. 1034-1040.

Boyles, M. (2011). Oklahoma State University. Herbicide foliar and residual injury. Cold temperatures symptoms. Symptoms similar to herbicide carryover injury. Available from: http://www.canola.okstate.edu/herbicides/foliarresidualinjury/index.htm

Brule-Babel, L., & Fowler, D. (1988). Genetic control of cold hardiness and vernalization requirement in wheat. *Crop Sci.* vol. 28, pp. 879–884.

Brzostowicz, A., & Barcikowska, V. (1987). Possibility of frost resistance testing of *Brassica napus* with the help of delayed luminescence intensity. *Cruciferae Newsletter*, pp. 12, 27.

Canola Connection. (2011). The Canola Council of Canada, markets & statistics. Available from: http://www.canolacouncil.org/acreageyields.aspx

Canola Council of Canada. (2011). Canola growers manual. Chapter 8 - Crop Establishment. Available from: http://www.canolacouncil.org/chapter8.aspx

Cardoza, V. & Stewart, N. (2004). *Brassica* biotechnology: Progress in cellular and molecular biology. In *Vitro Cell. & Dev. Biol. - Plant*, Vol. 40, pp. 542-551.

Carvallo, A., Pino, T., Jeknic, Z., Zou, C., Doherty, J., Shiu, H., Chen, H. & Thomashow, F. (2011). A comparison of the low temperature transcriptomes and CBF regulons of three plant species that differ in freezing tolerance: *Solanum commersonii, Solanum tuberosum,* and *Arabidopsis thaliana. Journal of Experimental Botany,* Vol. 629, No. 11, pp. 3807-3819.

Cebert, E. & Rufina, W. (2007). Genetic variation among winter canola cultivars to freezing temperatures during flowering and early seed formation. *Proceedings of ASA-CSSA-CSSA. International annual meetings,* Nov. 4-8, 2007. New Orleans, Louisiana, US.

Chachalis, D., & Smith, M. (2001). Hydrophobic-polymer application reduces imbibition rate and partially improves germination or emergence of soybean seedlings. *Seed Sci. Technol.,* Vol. 29, pp. 91-98.

Chengci, C., Grant, J., Karnes, N., David, W., Gregory, J., & Duane, J. (2005). Determining the feasibility of early seeding canola in the Northern Great Plains. *Agronomy Journal,* Vol. 97, No. 4, pp. 1252-1262.

Chen, L., Zhong, H., Ren, F., Guo, Q., Hu, P & Li, B. (2011). A novel cold-regulated gene, COR25, of *Brassica napus* is involved in plant response and tolerance to cold stress. *Plant Cell Rep.* Vol. 30, No. 4, pp. 463-71.

Cheung, F., Trick, M., Drou, N., Lim, P., Park, Y., Kwon, J., Kim, A., Scott, R., Pires, C., Paterson, H., Town, C., & Bancroft, I. (2009). Comparative analysis between homeologus genome segments of *Brassica napus* and its progenitor species reveals extensive sequence-level divergence. *The Plant Cell,* Vol. 21, pp. 1912-1928.

Chinnusamy, V., Zhu, K., & Sunkar, R. (2010). Gene Regulation during cold stress acclimation in plants. *Methods in Molecular Biology,* Vol. 639, No. 1, pp. 39-55.

Christian J. Willenborg, Robert H., Gulden, Eric N. Johnson, & Steven J. Shirtliffe. (2004). Germination Characteristics of Polymer-Coated Canola (*Brassica napus* L.) Seeds Subjected to Moisture Stress at Different Temperatures. *Agronomy Journal,* 96:3. 786-791.

Diaz, O., Gustafsson, M. & Astley, D. (1997). Effect of regeneration procedures on genetic diversity in *Brassica napus* and *B.rapa* as estimated by isozyme analysis. *Genetic Resources Crop Evolution,* Vol. 44, pp. 523-532.

Dong, C., Danyluk, J., Wilson, E., Pocock, T., Huner, A., & Sarhan, F. (2002). Cold-regulated cereal chloroplast late embryogenesis abundant-like proteins. Molecular characterization and functional analyses. *Plant Physiology,* Vol. 129, pp.1368-1381.

Downey, R.K. 1990. Canola: A quality *Brassica* oilseed. pp. 211-215. In: J. Janick and J.E. Simon (eds.), Advances in new crops. Timber Press, Portland, OR.

Epand, M., Shai, Y., Segrest, P., & Anantharamaiah, M. (1995). Mechanisms for the modulation of membrane bilayer properties by amphipathic helical peptides. *Biopolymers,* Vol. 37, pp. 319-338.

Fowler, D. B., Limin, A. E., Wang, S. Y. & Ward, R. W. (1996). Relationship between low-temperature tolerance and vernalization response in wheat and rye. *Can. J. Plant Sci.,* Vol. 76, pp.37-42.

Fowler, S., & Thomashow, J. (2002) *Arabidopsis* transcriptome profiling indicates that multiple regulatory pathways are activated during cold acclimation in addition to the CBF cold response. *Plant Cell,* Vol. 14, pp. 1675–90.

Fracheboud, Y., Haldimann, P., Leipner, J., & Stamp, P. (1999). Chlorophyll fluorescence as a selection tool for cold tolerance of photosynthesis in maize (*Zea mays* L.). *J. Exp. Bot.,* Vol. 50, pp. 1533-1540.

Gan, Y., Selles, F., & Angadi, S. (2001). Fall seeded canola: To coat or not to coat, that is the question. *Semi arid Prairie Agric. Res. Cent. News,* l. 17 Sept.

Gao, J., Allard, G., Byass, L., Flanagan, M. & Singh, J. (2002). Regulation and characterization of four *CBF* transcription factors from *Brassica napus. Plant Molecular Biology,* Vol. 49, pp. 459-471.

Garcia-C M., Figueros, J., Gomez, R., Townsend, R., & Schoper, J. (1998). Seed physiology, production & technology, pollen control during transgenic hybrid maize development in Mexico. *Crop Science,* Vol. 38, pp. 1597-1602.

Gilmour, J., Sebolt, M., Salazar, P., Everard, D., & Thomashow, F. (2000). Overexpression of the *Arabidopsis* CBF3 transcriptional activator mimics multiple biochemical changes associated with cold acclimation. *Plant Physiology,* Vol. 124, pp. 1854–1865.

Gosal, S., Wani, H., & Kang, S. (2009). Biotechnology and drought tolerance. *J. Crop Improv.,* Vol. 23, pp. 19–54.

Gray, R., Chauvin, P., Sarhan, F., & Huner, A. (1997). Cold acclimation and freezing tolerance. A complex interaction of light and temperature. *Plant Physiol.,* Vol. 114, pp. 467–474.

Grist, A., Firgaira, F., & Morley, A. (1993). Dinucleotide repeat polymorphisms isolated by the polymerase chain reaction. *Bio-Techniques,* Vol. 15, pp. 304–309.

Guy, C., Haskell, Dale., & Li, B. (1998). Association of proteins with the stress 70 molecular chaperones at low temperature: evidence for the existence of cold labile proteins in spinach. *Cryobiology,* Vol. 36. pp. 301-314.

Guy, C., & Li, B. (1998). The organization and evolution of the spinach stress 70 molecular chaperone gene family. *Plant Cell,* Vol. 10, pp. 539-556.

Haake, V., Cook, D., Riechmann, L., Pineda, O., Thomashow, F., & Zhang, Z. (2002). Transcription factor CBF4 is a regulator of drought adaptation in *Arabidopsis. Plant Physiology,* Vol. 130, pp. 639-648.

Hajela, K., Horvath, P., Gilmour, J., & Thomashow, F. (1990). Molecular cloning and expression of *cor* (cold-regulated) genes in *Arabidopsis thaliana. Plant Physiology,* Vol.93, pp. 1246-1252.

Hara, M., Terashima, S., Fukaya, T., & Kubol, T. (2003). Enhancement of cold tolerance and inhibition of lipid peroxidation by citrus dehydrin in transgenic tobacco. *Planta,* Vol. 217, pp. 290-298.

Harberd, J. (1969). A simple effective embryo culture technique for *Brassica. Euphitica,* Vol. 18, pp. 425-429.

Houde, M., Dallaire, S., Ndong, D., & Sarhan, F. (2004). Overexpression of the acidic dehydrin WCOR410 improves freezing tolerance in transgenic strawberry leaves. *Plant Biotechnology Journal,* Vol. 2, pp. 381-387.

Hsieh, H., Lee, T., Yang, T., Chiu, H., Charng, Y., Wang, C., & Chan, T. (2002). Heterology expression of the *Arabidopsis* C-repeat/dehydration response element binding factor 1 gene confers elevated tolerance to chilling and oxidative stresses in transgenic tomato. *Plant Physiol.*, Vol. 129, pp. 1086–1094.

Iturriaga, G., Scheinder, K., Salamini, F. & Bartels, D. (1992). Expression of desiccation-related proteins from the resurrection plant *Craterostigma plantagineum* in transgenic tobacco. *Plant Molecular Biology*, Vol. 20, pp. 555-558.

Jaglo, O., Kirsten, R., Gilmour, J., Zarka, G., Oliver, S., & Thomashow, F. (1998). *Arabidopsis* CBF1 overexpression induces *COR* genes and enhances freezing tolerance. *Science*, Vol. 280, pp. 104-106.

Jaglo, R., Kleff, S., Amundsen, L., Zhang, X., Haake, V., Zhang, Z., Deits, T., & Thomashow, F. (2001). Components of the *Arabidopsis* C-Repeat/Dehydration-Responsive Element Binding Factor Cold-Response Pathway are Conserved in *Brassica napus* and Other Plant Species. *Plant Physiology*, Vol. 127, pp. 910–917.

Jaglo-Ottosen, R., Gilmour, J., Zarka, G., Schabenberger, O., & Thomashow, F. (1998). *Arabidopsis* CBF1 overexpression induces COR genes and enhances freezing tolerance. *Science*, Vol. 280, pp. 104–106.

Jiang, C., Iu, B., & Singh, J. (1996). Requirement of a CCGAC cis-acting element for cold induction of the *BNll5* gene from winter *Brassica napus*. *Plant Molecular Biology*, Vol. 30, pp. 679–684.

Johnson, L., McKay, K., Schneiter, A., Hanson, B., & Schatz, B. (1995). Influence of planting date on canola and crambe production. *J. Prod. Agric.*, Vol. 8, pp. 594-599.

Kacperska A. 1983. Mechanism of cold acclimation in winter rape plant. *Proc. 6th Congress Inter. canola*, Paris, 17-19.05.1983, t. 1: 78-82.

Kaye, C., Neven, L., Hogig, A., Li, B., Haskell, D., & Guy, C. (1998). Characterization of a gene for spinach CAP160 and expression of two spinach cold-acclimation proteins in tobacco. *Plant Physiology*, Vol. 116, pp. 1367-1377.

Kim, S., Park, J., Kwak, J., Kim, O., Kim, Y., Song, J., Jang, B., Jung, H., & Kang, H. (2007). Cold shock domain proteins and glycine-rich RNA-binding proteins from *Arabidopsis thaliana* can promote the cold adaptation process in *Escherichia coli*. *Nucleic Acids Res.*, Vol.35, pp. 506-516.

Kirkland, J., & Johnson, E. (2000). Alternative seeding dates (fall and April) affect *Brassica napus* canola yield and quality. *Can. J. Plant Sci.*, Vol. 80, pp. 713-719.

Knight, H., Veale, L., Warren, J., & Knight, R. (1999). The sfr6 mutation in *Arabidopsis* suppresses low temperature induction of genes dependent on the CRT/DRE sequence motif. *Plant Cell*, Vol. 11, pp. 875–886.

Kott, S., Polsoni, L., & Beversdorf, D. (1988). Cytological aspects of isolated microspore culture of *Brassica napus*. *Can. J. Botany*, Vol. 66, pp. 1658-1664.

Kreps, A., Wu, Y., Chang, S., Zhu, T., Wang, X., & Harper, F. (2002) Transcriptome changes for *Arabidopsis* in response to salt, osmotic, and cold stress. *Plant Physiol.*, Vol. 130, pp. 2129-2141.

Krishan, P., Sacco, M., Cherutti, F., & Hill, S. (1995). Cold-Induced accumulation of hsp90 transcripts in *Brassica napus*. *Plant Physiol.*, Vol. 107, pp. 915-923

Kumar, N., & Bhatt, P. (2006). Transgenics: An emerging approach for cold tolerance to enhance vegetables production in high altitude areas. *Indian J. Crop Sci.*, Vol. 1, pp. 8-12.

Kutter, C., Schob, H., Stadler, M., Meins, F., & Ammour, S. (2007). MicroRNA-mediated regulation of stomatal development in *Arabidopsis*. *Plant Cell*, Vol. 19, No. 8, pp. 2417-2429.

Landry, S., Hubert, N., Etoh, T., Harada, J., & Lincoln, S. (1991). A genetic map for *Brassica napus* based on restriction fragment length polymorphisms detected with expressed DNA sequences. *Genome*, Vol. 34, pp. 543–552.

Laroche, A., Geng, X., & Singh, J. (1992). Differences of freezing tolerance and vernalization responses in cruciferae exposed to low temperature. *Plant Cell. Environ.*, Vol. 15, pp. 439-445.

Lata, C., & Prasad, M. (2011). Role of DREBs in regulation of abiotic stress responses in plants. *Journal of Experimental Botany*, Vol. 62, (14): 4731-4748.

Lee, C., Lim, H., Kim, A., Lee, I., Kim, S., Jin, M., Kwon, J., Mun, H., Kim, K., Kim, U., Hur, Y., & Park, S. (2008). Transcriptome analysis in *Brassica rapa* under the abiotic stresses using *Brassica* 24K oligo microarray. *Mol Cells*, Vol. 26, No. 6, pp. 595-605.

Lee, H., Xiong, L., Gong, Z., Ishitani, M., Stevenson, B., & Zhu, K. (2001). The *Arabidopsis* HOS1 gene negatively regulates cold signal transduction and encodes a ZINC finger protein that displays cold-regulated nucleo-cytoplasmic partitioning. *Genes Dev.*, Vol. 15, pp. 912–924.

Livingston, J., & de Jong, E. (1990). Matric and osmotic potential effects on seedling emergence at different temperatures. *Agron. J.*, Vol. 82, pp. 995–998.

Livingston, P., & Henson, A. (1998). Apoplastic sugars, fructans, fructan exohydrolase, and invertase in winter oat: responses to second-phase cold hardening. *Plant Physiology*, Vol. 116, pp. 403-408.

Madakadze, C., Stewart, K., Madakadze, R., & Smith, D. (2003). Base temperature for seedling growth and their correlation with chilling sensitivity for warm-season grasses. *Crop Sci.*, Vol. 43, pp. 874-878.

Markowski, A., & Rapacz, M. (1994). Comparison of vernalization requirement and frost resistance of winter rape lines derived from doubled haploid. *J. Agric. Crop Sci.*, Vol. 173, pp. 184-192.

Mieczyslaw, K. (1999). Cytochemical localization of phenolic compounds in columella cells of the root cap in Seeds of *Brassica napus*–Changes in the localization of phenolic compounds during germination. *Annals of Botany*, Vol. 84, pp.135–143.

Miura, K., Jin, B., Lee, J., Yoo, Y., Strim, V., Miura, T., Ashworth, N., Bressan, A., Yun, J., & Hasegawa, M. (2007). SIZ1-mediated sumoylation of ICE1 controls *CBF3/DREB1A* expression and freezing tolerance in *Arabidopsis*. *Plant Cell*, Vol. 19, pp. 1403-1414.

Mohaptra, D., & Bajaj, S. (1987). Interspecific hybridization in *Brassica juncea* x *B. hirta* using embryo rescue. *Euphitica*, Vol. 36, pp. 321-326.

Morrison, J., McVetty, P., & Shaykewich, C. (1989). The determination and verification of a baseline temperature for the growth of Westar summer rape. *Can. J. Plant Sci.*, Vol. 69, pp. 455-464.

Murata, N., Ishizaki-Nishizawa, O., Higashi, S., Hayashi, S., Tasaka, Y., & Nishida, I. (1992). Genetically engineered alteration in the chilling sensitivity of plants. *Nature*, 1992, 356, 710-713.

Murray, B., Cape, J., & Fowler, D. (1989). Quantification of frost damage in plant tissue by rates of electrolyte leakage. *New Phytol.*, Vol. 113, pp. 307-311.

Nakashima, K., & Yamaguchi-Shinozaki, K. (2006). Regulons involved in osmotic stress-responsive and cold stress-responsive gene expression in plants. *Physiologia Plantarum*, Vol. 126, No. 1, pp. 62-71.

National Agricultural Statistic Service. (2011). U.S. and state level data. Agricultural Statistics Board, United States Department of Agriculture (USDA). Available from: http://usda.mannlib.cornell.edu/usda/current/Acre/Acre-06-30-2011.pdf

Nishi, S., Kawata, J., & Toda, M. (1959). On the Breeding of interspecific hybrids of *Brassica* through the application of embryo culture techniques. *Jap. J. of Breeding*, Vol. 8, pp. 215-222.

Nykiforuk, L., & Johnson-Flanagan, A. (1994). Germination and early seedling development under low temperature in canola. *Crop Sci.*, Vol. 34, pp. 1047-1054.

Ofori, A., Becker, H., & Kopisch-Obuch, F. (2008). Effect of crop improvement on genetic diversity in oilseed *Brassica rapa* (turnip-rape) cultivars, detected by SSR markers. *Journal of Applied Genetics*, Vol. 49, No. 3, pp. 207-212.

Pennycooke, C., Jones, L., & Stushnoff, C. (2003). Down-regulating α-Galactosidase enhances freezing tolerance in transgenic Petunia. *Plant Physiol.*, Vol. 133, pp. 901-909.

Puhakainen, T., Hess, W., Makela, P., Svensson, J., Heino, P., & Palva, T. (2004). Overexpression of multiple dehydrin genes enhances tolerance to freezing stress in *Arabidopsis. Plant Molecular Biology*, Vol. 54, pp. 743-753.

Quazi, H. (1988). Interspecific hybrids between *Brassica napus* L. and *B. oleracea* L. developed by embryo culture. *Theor. Appl. Genet.*, Vol. 75, pp. 309-318.

Rapacz, M. (2002). Cold-deacclimation of oilseed rape (*Brassica napus* var. *oleifera*) in response to fluctuating temperatures and photoperiod. *Annals of Botany*, Vol. 89, pp. 543–549.

Rapacz, M., & Markowski, A. (1999). Winter hardiness, frost resistance and vernalization requirement of European winter oilseed rape (*Brassica napus* var *oleifera*) cultivars within the last 20 years. *J. Agric. Crop Sci.* Vol. 183, pp. 243–253.

Rife, C., Heer, W., Janssen, K., Long, J., Evans, P., Aiken, R., Witt, M., Auld, D., Bacon, R., Baltensperger, D., Beans, B., Bishnoi, U., Bhardwaj, H., Bordovsky, D., Christmas, E., Copeland, L., Ivy, R., Johnson, D., Kelly, J., Morris, C., Nelson, L., Raymer, P., Saunders, R., Schmidt, M., Starner, D., & Weibold, W. (2001). 2000 national winter canola variety trial departmental report. Kansas Agric. Exp. Stn. Manhattan.

Rife, L., & Zeinali, H. (2003). Cold tolerance in oilseed rape over varying acclimation durations. *Crop Science*, Vol. 43, No. 1, pp. 96-100.

Rizza, D., Pagani, A., Stanca, M., & Cattivelli, L. (2001). Use of chlorophyll fluorescence to evaluate the cold acclimation and freezing tolerance of winter and spring oats. Plant *Breeding*, Vol. 120, pp. 389-396.

Ryschka, U., Marthe, F., Klocke, E., Schumann, G., & Zhao, H. (2007). Culture and fusion of pollen protoplasts of *Brassica oleracea* L. var. *italica* with haploid mesophyll protoplasts of *B. rapa* L. ssp. *Pekinensis. Protoplasma*, Vol. 231, No. 1-2, pp. 89-97.

Sacristan, D., & Gerdemann, M. (1986). Different behavior of *Brassica juncea* and *B. Carinata* as sources of *Phoma lingam* resistance in experiments of interspecific transfer to *B. Napus. Plant Breeding*, Vol. 97, pp. 304-314.

Sangwan, V., Foulds, I., Singh, J., & Dhindsa, R. (2001). Cold-activation of *Brassica napus* *BN115* promoter is mediated by structural changes in membranes and cytoskeleton, and requires Ca^{2+} influx. *The Plant Journal*, Vol. 27, No. 1, pp. 1-12.

Savitch V., Allard, G., Seki, M., Robert, S., Tinker, A., Huner, A., Shinozaki, K., & Singh, J. (2005). The Effect of Overexpression of Two *Brassica* CBF/DREB1-like Transcription Factors on Photosynthetic Capacity and Freezing Tolerance in *Brassica napus*. *Plant Cell Physiol.*, Vol. 46, No. 9, pp. 1525–1539.

Seki, M., Narusaka, M., Abe, H., Kasuga, M., Yamaguchi-Shinozaki, K., Carninci, P., Hayashizaki, Y., & Shinozaki, K. (2001) Monitoring the expression pattern of the 1300 *Arabidopsis* genes under drought and cold stresses by using full length cDNA microarrays. *Plant Cell*, Vol. 13, pp. 61–72.

Seki, M., Narusaka, M., Ishida, J., Nanjo, T., Fujita, M., Oono, Y., Kamiya, A., Nakajima, M., Enju, A., Sakurai, T., Satou, M., Akiyama, K., Satou, M., Taji, T., Yamaguchi-Shinozaki, K., Carninci, P., Kawai, J., Hayashizaki, Y., & Shinozaki, K. (2002). Monitoring the expression profiles of 7000 *Arabidopsis* genes under drought, cold, and high-salinity stresses using a full-length cDNA microarray. *Plant J*. Vol. 31, pp. 279–292.

Shinozaki, K., Yamaguchi-Shinozaki, K., & Seki, M. (2003). Regulatory network of gene expression in the drought and cold stress responses. *Curr Opin Plant Biol.*, Vol. 6, pp. 410-417.

Singh, S. K., V. G. Kakani, D. Brand, B. Baldwin, & K. R. Reddy. (2008). Assessment of cold and heat tolerance of winter-grown canola *(Brassica napus* L.) cultivars by Pollen-based parameters. *J. Agronomy & Crop Science*, Vol. 194, No. 3, pp. 225-236.

Slocum, K., Figdore, S., Kennard, W., Suzuki, J., & Osborn, T. (1990). Linkage arrangement of restriction fragment length polymorphisms in *Brassica oleracea*. *Theor. Appl. Genet.*, Vol. 80, pp. 57 – 64.

Snowdon, J., & Friedt, W. (2004). Molecular markers in *Brassica* oilseed breeding: current status and future possibilities. *Plant Breeding*, Vol. 123, pp. 1-8.

Song, M., Suzuki, J., Slocum, M., Williams, P., & Osborn, T. (1991). A linkage map of *Brassica rapa* (syn. *campestris*) based on restriction fragment length polymorphisms. *Theor. Appl. Genet.*, Vol. 82, pp. 296 – 304.

Sovero, M. (1993). Rapeseed, a new oilseed crop for the United States, In: *New crops*, J. Janick and J.E. Simon (eds.), pp. 302-307, Wiley, New York.

Stefanowska, M., Kuras, M., & Kacperska, A. (2002) Low temperature induced modifications in cell ultrastructure and localization of phenolics in winter oilseed rape *(Brassica napus* L. var *oleifera* L.) leaves. *Ann. Bot.*, Vol. 90, pp. 637–64.

Stefanowska, M., Kuras, M., Kubacka-Zebalska, M., & Kacperska, A. (1999) Low temperature affects pattern of leaf growth and structure of cell walls in winter oilseed rape *(Brassica napus* L. var *oleifera* L.). *Ann. Bot.*, Vol. 84, pp. 313–319.

Stockinger, J., Gilmour, J. & Thomashow, M. (1997). *Arabidopsis thaliana* CBF1 encodes an AP2 domain-containing transcription activator that binds to the C-repeat/DRE, a *cis* acting DNA regulatory element that stimulates transcription in response to low temperature and water deficit. *Proceedings of National Academy of Sciences*, Vol .94, pp. 1035–1040.

Strand, Å., Hurry, V., Henkes, S., Huner, N., Gustafsson, P., Gardestrom, P., & Stitt, M. (1999) Acclimation of *Arabidopsis* leaves developing at low temperatures. Increasing cytoplasmic volume accompanies increased activities of enzymes in the Calvin cycle and in the sucrose-biosynthesis pathway. *Plant Physiol.*, Vol. 119, pp. 1387–1397.

Su, F., Wang, C., Hsieh, H., Lu, A., Tseng, H., & Yu, M. (2010). A novel MYBS3-dependent pathway confers cold tolerance in rice. *Plant Physiol.*, Vol. 153, pp. 145-158.

Taji, T., Ohsumi, C., Iuchi, S., Seki, M., Kasuga, M., Kobayashi, M., Yamaguchi-Shinozaki, K., & Shinozaki, K. (2002). Important roles of drought- and cold-inducible genes for galactinol synthase in stress tolerance in *Arabidopsis thaliana*. *The Plant Journal*, Vol. 29, No. 4, pp. 417-426.

Teutonico, A., Palta P., & Osborn, C. (1993). In vitro freezing tolerance in relation to winter survival of rapeseed cultivars. *Crop Sci.*, Vol. 33, pp. 103-107.

The *Brassica rapa* Genome Sequencing Project Consortium. 2011. Xiaowu Wang, Hanzhong Wang, Jun Wang, Rifei Sun, Jian Wu, Shengyi Liu, Yinqi Bai, Jeong-Hwan Mun, Ian Bancroft, Feng Cheng, Sanwen Huang, Xixiang Li, Wei Hua, Junyi Wang, Xiyin Wang, Michael Freeling, J Chris Pires, Andrew H Paterson, Boulos Chalhoub, Bo Wang, Alice Hayward, Andrew G Sharpe, Beom-Seok Park, Bernd Weisshaar, Binghang Liu, Bo Li, Bo Liu, Chaobo Tong, Chi Song, Christopher Duran, Chunfang Peng, Chunyu Geng, Chushin Koh, Chuyu Lin, David Edwards, Desheng Mu, Di Shen, Eleni Soumpourou, Fei Li, Fiona Fraser, Gavin Conant, Gilles Lassalle, Graham J King, Guusje Bonnema, Haibao Tang, Haiping Wang, Harry Belcram, Heling Zhou, Hideki Hirakawa, Hiroshi Abe, Hui Guo, Hui Wang, Huizhe Jin, Isobel A P Parkin, Jacqueline Batley, Jeong-Sun Kim, Jérémy Just, Jianwen Li, Jiaohui Xu, Jie Deng, Jin A Kim, Jingping Li, Jingyin Yu, Jinling Meng, Jinpeng Wang, Jiumeng Min, Julie Poulain, Jun Wang, Katsunori Hatakeyama, Kui Wu, Li Wang, Lu Fang, Martin Trick, Matthew G Links, Meixia Zhao, Mina Jin, Nirala Ramchiary, Nizar Drou, Paul J Berkman, Qingle Cai, Quanfei Huang, Ruiqiang Li, Satoshi Tabata, Shifeng Cheng, Shu Zhang, Shujiang Zhang, Shunmou Huang, Shusei Sato, Silong Sun, Soo-Jin Kwon, Su-Ryun Choi, Tae-Ho Lee, Wei Fan, Xiang Zhao, Xu Tan, Xun Xu, Yan Wang, Yang Qiu, Ye Yin, Yingrui Li, Yongchen Du, Yongcui Liao, Yongpyo Lim, Yoshihiro Narusaka, Yupeng Wang, Zhenyi Wang, Zhenyu Li, Zhiwen Wang, Zhiyong Xiong, & Zhonghua Zhang. The genome of the mesopolyploid crop species *Brassica rapa*. *Nature Genetics*, Volume: 43, Pages: 1035–1039.

Thomashow, F. (2001). So what's new in the field of plant cold acclimation? Lots! *Plant Physiol.*, Vol. 125, pp. 89-93.

Thomashow, M. (1999). Plant cold acclimation: freezing tolerance genes and regulatory mechanisms. Annu Rev Plant Physiol Plant Mol Biol., Vol. 50, pp. 571-599.

Trischuk, G., Schilling, B., Wisniewsky, M., & Gusta, L. (2006). Physiology and Molecular Biology of Stress Tolerance in Plants: in Freezing Stress, In: *Systems Biology to Study Cold Tolerance* K. V. Mahava Rao, A. S. Raghavendra and K. Janardham (eds), pp. 131-155, Springer//, Nethelands.

Ulrich, D. (2005). Mutual interactions between the SUMO and ubiquitin systems: A plea of no contest. *Trends in Cell Biology*, Vol. 15, pp. 525-532.

Nagaharu, U. (1935). Genome analysis in *Brassica* with special reference to the experimental formation of B. napus and peculiar mode of fertilization. Japan. J. Bot 7: 389–452.

Valdes, M., & Bradford, K. (1987). Effects of seed coating and osmotic priming on the germination of lettuce seeds. *J. Am. Soc. Hortic. Sci.* Vol. 112, pp. 153–156.

Vigil, F., Anderson, R., & Beard, W. (1997). Base temperature and growing-degree-hour requirements for the emergence of canola. *Crop Sci.* Vol. 37, pp. 844-849.

Vos, P., Hogers, R., Sleeker, M., Reijans, M., Lee, T., Homes, M., Freiters, A., Pot, J., Peleman, J., Kuiper, M., & Zabeau, M. (1995). AFLP: a new concept for DNA fingerprinting. *Nucl. Acids Res.*, Vol. 23, pp. 4407 – 4414.

Wanner, A., & Junttila, O. (1999) Cold-induced freezing tolerance in *Arabidopsis. Plant Physiol.* Vol. 120, pp. 391–399.

Wani, H., & Gosal, S. (2011). Introduction of OsglyII gene into Indica rice through particle bombardment for increased salinity tolerance. *Biol. Plant.* Vol. 55, No. 3, pp. 536-540.

Wani, H., Sandhu, S., & Gosal, S. (2008). Genetic engineering of crop plants for abiotic stress tolerance. In: *Advanced Topics in Plant Biotechnology and Plant Biology,* Malik CP, Kaur B, Wadhwani C, editors, pp. 149–183. MD Publications, New Delhi. .

Willenborg, J., Gulden, R., Johnson., E., & Shirtliffe, S. (2004).Germination characteristics of polymer-coated canola (*Brassica napus* L.) seeds subjected to moisture stress at different temperatures. *Agron. J.,* Vol. 96, pp. 786-791.

Williams, K., Kubelik, A., Livak, K., Rafalski, J., & Tingey, S. (1990). DNA polymorphisms amplified by arbitrary primers are useful as genetic markers. *Nucl. Acids Res.,* Vol. 18, pp. 6531 – 6535.

Yang, J., Kim, S., Lim, B., Kwon, J., Kim, A., Jin, M., Park, Y., Lim, H., Kim, H., Kim, H., Lim, P., & Park, S. (2005). The Korean *Brassica* genome project: a glimpse of the *Brassica* genome based on comparative genome analysis with *Arabidopsis. Comp Funct Genomics,* Vol. 6, No. 3, pp. 138-146.

Yamaguchi-Shinozaki, K., & Shinozaki, K. (2006). Transcriptional regulatory networks in cellular responses and tolerance to dehydration and cold stresses. *Annu. Rev. Plant Biol.,* Vol. 57, pp. 781-803.

Yamaguchi-Shinozaki, K., & Shinozaki, K. (1994). A Novel cis-Acting Element in an *Arabidopsis* Gene Is Involved in Responsiveness to Drought, Low-Temperature, or High-Salt Stress. *The Plant Cell,* Vol. 692, pp. 251-264.

Zaychuk, S., & Enders, N. (2001). Water soluble, freeze sensitive seed coatings. U.S. Patent 6 230 438. Date issued: 15 May.

Zhang, Q., Zhou, W., Gu, H., Song, W., & Momoh, E. (2003). Plant regeneration from the hybridization of *Brassica juncea* and *B. napus* through embryo culture. *J. Agron. Crop Sci.,* Vol. 189, pp. 347-350.

Zhang, X., Fowler, G., Cheng, H., Lou, Y., Rhee, S., Stockinger, E., & Thomashow, M. (2004b). Freezing-sensitive tomato has a functional CBF cold response pathway, but a CBF regulon that differs from that of freezing-tolerant *Arabidopsis. Plant J.,* Vol. 39, pp. 905-919.

Zhang, Z., Creelman, R., & Zhu, K. (2004a). From Laboratory to Field. Using information from *Arabidopsis* to Engineer Salt, Cold, and Drought Tolerance in Crops. *Plant Physiology,* Vol. 135, pp. 615-621.

Zheng, H., Wilen, R., Slinkard, A., & Gusta, L. (1994). Enhancement of canola seed germination and seedling emergence at low temperature by priming. *Crop Sci.,* Vol. 34, pp. 1589-1593.

Zhu, Q., Zhang, T., Tang, R., Lv, D., Wang, Q., Yang L., & Zhang, X. (2009). Molecular characterization of *ThIPK2*, an inositol polyphosphate kinase gene homolog from *Thellungiella halophila*, and its heterologous expression to improve abiotic stress tolerance in *Brassica napus*. *Physiologia Plantarum*, Vol. 136, No. 4, pp. 407-425.

Zietkiewicz, E., Rafalski, A., & Labuda, D. (1994). Genome fingerprinting by simple sequence repeat (SSR)-anchored polymerase chain reaction amplification. *Genomics*, Vol. 20, pp. 176–183.

Sesame Seed

T. Y. Tunde-Akintunde[1], M. O. Oke[1] and B. O. Akintunde[2]
[1]Department of Food Science and Engineering,
Ladoke Akintola University of Technology, Ogbomoso,
[2]Federal College of Agriculture, I.A.R. and T., P.M.B. 5029, Ibadan, Oyo State,
Nigeria

1. Introduction

Sesame (*Sesamum indicum* L.), otherwise known as sesamum or benniseed, member of the family *Pedaliaceae*, is one of the most ancient oilseeds crop known to mankind. Sesame plays an important role in human nutrition. Most of the sesame seeds are used for oil extraction and the rest are used for edible purposes (El Khier et al, 2008). Sesame is grown primarily for its oil-rich seeds. Before seeds were appreciated for their ability to add nutty flavour or garnish foods, they were primarily used for oil and wine (Ghandi, 2009). After the extraction of oil, the cake is mostly used for livestock feed or often as manure. Its colour varies from cream-white to charcoal-black but it is mainly white or black. Other colours of some sesame seed varieties include, yellow, red or brown (Naturland, 2002). In Nigeria, the notable colours for sesame seed are white, yellow and black (Fariku et al., 2007). The lighter varieties of sesame which are considered to be of higher quality are generally more valued in the West and Middle East, while both the pale and black varieties are prized in the Far East. (www.wikepedia-sesame). There are numerous varieties and ecotypes of sesame adapted to various ecological conditions (Nzioku et al., 2010).

The major world producers include India, Sudan, China and Burma (who contribute about 60% of the total world production) (El Khier et al, 2008). It is also one of main commercial crops in Nigeria, Sudan and Ethiopia (www. nutrition and you). Sesame is an important crop to Nigerian agriculture: it is quite extensively cultivated especially in Northern Nigeria. It yields in relatively poor climatic conditions, and it is widely used within Nigeria. Moreso, it is an important component of Nigeria's agricultural exports (Chemonics, 2002).

Sesame seed is rich in fat, protein, carbohydrates, fibre and some minerals. The oil seed is renowned for its stability because it strongly resists oxidative rancidity even after long exposure to air (Global AgriSystems, 2010). The oil fraction shows a remarkable stability to oxidation. This could be attributed to endogenous antioxidants namely lignins and tocopherols (Elleuch et al., 2007; Lee et al., 2008).

The seed is rich in protein and the protein has disable amino acid profile with good nutritional value similar to soybean (NAERLS, 2010). The chemical composition of sesame shows that the seed is an important source of oil (44-58%), protein (18-25%), carbohydrate (~13.5%) and ash (~5%) (Borchani et al., 2010). Sesame seed is approximately 50 percent oil (out of which 35% is monounsaturated fatty acids and 44% polyunsaturated fatty acids) and 45 percent meal (out of which 20% is protein) (Ghandi,2009; Hansen, 2011).

Sesame seeds are small, almost oblate in shape and have a mild and delicious aroma and taste. Sesame seed is used whole in cooking and also yields sesame oil (www.wikepedia,sesame; Hansen, 2011). It has a rich nutty flavour (although such heating damages their healthful polyunsaturated fats) and is used mainly as a food ingredient in whole, broken, crushed, shelled, powdered and paste forms. Its use is country based, in the US; it is used as some form of whole seed product for the confection and baking industries. A small percentage percent of total production is however processed into oil, meal or flour (Hansen, 2011). In Nigeria, the seeds are consumed fresh, dried, fried or when blended with sugar. It is also used as a paste in some local soups. (Fariku et al., 2007).

Sesame seeds are not only used for culinary purposes but also in traditional medicines for their nutritive, preventive and curative properties. Its oil seeds are sources for some phyto-nutrients such as omega-6 fatty acids, flavonoid phenolic anti-oxidants, vitamins and dietary fiber with potent anti-cancer as well as health promoting properties (www.nutrition-and-you). Sesame oil is an edible vegetable oil derived from sesame seeds used in various countries. It is used as a cooking oil in South India and Asia and often as a flavor enhancer in Chinese, Japanese, Korean, and to a lesser extent Southern Asia cuisine (www.wikepedia-sesame.oil).

It is stable and free from undesirable nutrition or flavor component. Beniseed oil has a natural oxidant which prevents aging and is vital for the production of liver cells (NAERLS, 2010; Weiss, 2000). The oil also contains oleic (35.9-47%), linoleic (35.6-47.6), palmitic (8.7-13.8%), stearic (2.1-6.4%), as well as arachidic acids (0.1-0.7%) (Elleuch et al., 2007; Borchani et al., 2010).

2. History

Sesame seeds are the seeds of the tropical annual Sesamum indicum. The species has a long history of cultivation, mostly for its yield of oil. The oil plant has been grown since the beginning of arable cultivation, and originates from the dry bush savannah of tropical Africa, and spread from there to India and China, where it is still widely cultivated. (Naturland, 2002). The original area of domestication of sesame is obscure but it seems likely to have first been brought into cultivation in Asia or India (www.nigeriamarkets.org). Archeological records indicate that it has been known and used in India for more than 5,000 years and is recorded as a crop in Babylon and Assyria some 4,000 years ago (Borchani et al., 2010).

Sesame was cultivated during the Indus valley civilization and was the main oil crop. It was probably exported to Mesopotamia around 2500 BCE and was known in Akkadian and Sumerian as 'ellu'. Prior to 600 BC, the Assyrians used sesame oil as a food, salve, and medication, primarily by the rich, as the difficulty of obtaining it made it expensive. Hindus used it in votive lamps and considered the oil sacred.

3. Production of sesame seed

Global production of sesame seed is estimated by FAO at 3.15 mn tonnes per year (2001) having risen from 1.4 mn tonnes in the early 1960's. However only a small proportion of the global sesame harvest enters international trade. For the most part, the oil is expressed locally and used locally for cooking or the seeds themselves are eaten, particularly after being fried.

Sesame is grown in many parts of the world on over 5 million acres (20,000 km²). The largest producer of the crop in 2007 was India, China, Myanmar, Sudan, Ethiopia, Uganda and Nigeria. Seventy percent of the world's sesame crop is grown in Asia, with Africa growing 26% (Hansen, 2011)

The largest producers are China and India, each with an annual harvest around 750,000 tonnes followed by Myanmar (425,000 tonnes) and Sudan (300,000 tonnes). These figures are only rough estimates of the situation as sesame is a smallholder crop and much of the harvest is consumed locally, without record of the internal trade and domestic processing (Figure 1). Nigeria has a great market potential for sesame seed production for domestic and export markets noting that the production figures of the commodity has been on a steady increase since 1980, reaching 67000 MT by 1997 and was estimated to reach 139, 000 MT by the year 2010, according to the federal ministry of agriculture and natural resources (Joseph, 2009). This is agreement with the 2008 annual report of the Central Bank of Nigeria which states there has been a rise in production of sesame seed from 98,000,000 to 152,000,000 kg from 2003 to 2007 (CBN, 2009).

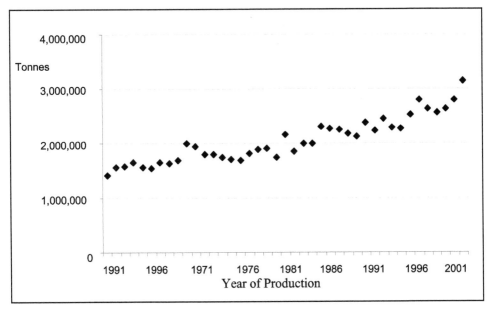

Fig. 1. World Production of Sesame Seed

Out of the estimated 3.5million hectares of Nigeria's arable land suitable for the growth of sesame seed, only 300,000 is currently used for the crop. However average yield of crop is about 300kg/ha which is 4 times lower than the average yield of other seed crops eg groundnut and soybeans. In major production zones in the country, it is used in traditional food recipes and snacks rather than for export purposes (NAERLS, 2010). Nigeria was the largest supplier to the Japanese market, the world's largest import market for sesame (Chemonics, 2002). Thus, the potentials for beniseed production in Nigeria is high since Japan, as well as Taiwan and Korea, generate global demand and offer opportunity for

Nigerian growers. Nigeria has a 6% share of the $600 million global market for sesame seed (Nigeria's Harvest, 2009).

Sesame was widely grown in Middle Belt, Northern and Central Nigeria as a minor crop initially in 1974 when it became a major cash crop in many Northern States eg Benue, Kogi, Gombe, Jigawa, Kano, Nasarawa, Katsina, Plateau, Yobe and Federal capital Territory (NAERLS, 2010). Sesame is commonly grown by smallholder farmers. The major producing areas in order of priority are Nasarawa, Jigawa and Benue States. Other important areas of production are found in Yobe, Niger, Kano, Katsina, Kogi, Gombe and Plateau States (Figure 2). Harvesting begins in late December and continues through July. Each producing area has only one season.

There are 2 types of sesame produced in Nigeria
1. White/raw = Food-grade used in bakery industry. 98-100% whitest grade seeds
2. Brown/mixed = Primarily oil-grade

The White (Food Grade) seed is grown around the towns of Keffi, Lafia/Makurdi, Doma, and in Nassarawa, Taraba, and Benue States. It is easier to sort and the Fumani/Denin people consume sesame locally. The Brown/mixed grows in the North, in Kano State and in Jigawa State near Hadejia, and somewhat in the southern part of Katsina State. There is some local consumption of the brown grade, but not much. The brown can be upgraded to Food Grade through bleaching, as discussed earlier (Chemonics, 2002). Several varieties of sesame are cultivated in Nigeria. The basic agronomic characteristics of some of the varieties are shown in Table 1 below.

Variety	Days to Maturity	Seed Color	Seed Size (mm)	Oil Content (%)	Potential Yield (kg/ha)
NCRI BEN-01M	102-115	White	3	45	1000
NCRI BEN-02M	102-115	Light	3	45	750
NCRI BEN-032	125-140	Brown	2	40	600
E-8	90	White	3.6	50	1000
Yandev-55	125	Light	2.5	45	600

Table 1. Characteristics of Sesame Grown in Nigeria

Growth conditions

Sesame has important agricultural attributes: it is adapted to tropical and temperate conditions, grows well on stored soil moisture with minimal irrigation or rainfall, can produce good yields under high temperatures, and its grain has a high value (Bennet, 2011). Sesame is found growing in most tropical, sub tropical and southern temperate areas of the world (Ghandi, 2009). However it is now cultivated around the dry tropics between the latitudes of 40° N and S. It is scarcely cultivated in the USA or Europe, not only because of climate but also because of the low returns per unit area (Chemonics, 2002). It has been reported to be a typical crop for small farmers in the developing countries (Bennet, 2011).

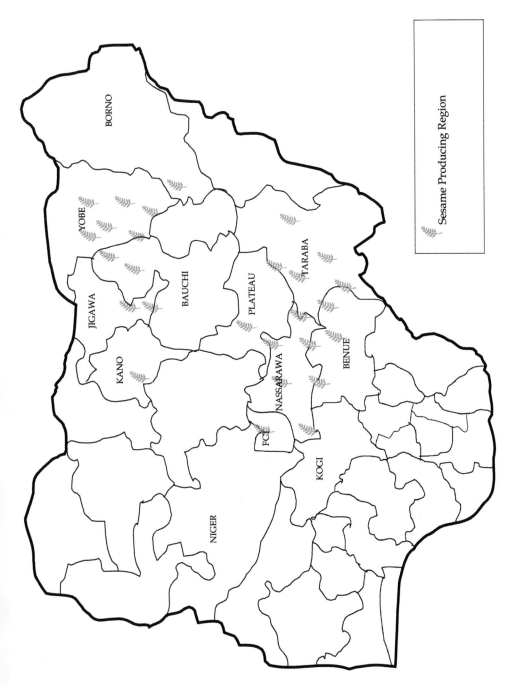

Fig. 2. Map of Nigeria Showing Sesame Producing Regions

This is because it has deep roots and are well adapted to withstand dry conditions. It will grow on relatively poor soils in climates generally unsuitable for other crops, and so it is widely valued for its nutritional and financial yield from otherwise inclement areas. It is well suited to smallholder farming with a relatively short harvest cycle of 90 –140 days allowing other crops to be grown in the field (Nigeria's Harvest, 2009; Naturland, 2002, Chemonics, 2002) and is often intercropped with other grains. This makes it favourable to Nigerian farmers and production can thus be sustained by small scale farmers with minimum management with an average yield of 700 kilograms per hectare (Nigeria's Harvest, 2009).

Sesame world production areas have remained generally stable over the years, but in some countries the crop is being marginalized (Bennet, 2011). Competition from more remunerative crops and a shortage of labour have pushed sesame to the less fertile fields and to areas of higher risk. Left unchecked, sesame production may decrease in the foreseeable future.

Sesame cultivation can be carried out on a wide range of soils but optimum are well-drained, loose, fertile and sandy alluvial soils that have a pH value between 5.4 and 6.75. Very low pH values have a drastic effect on growth, whereas some varieties can tolerate a pH value up to 8 (Naturland, 2002). Good drainage is crucial, as sesame is very susceptible to short periods of waterlogging. Sesame is intolerant of very acidic or saline soils (Bennet, 2011). The total amount of water required to grow sesame crop ranges from 600 to 1000 mm, depending on the cultivar and the climatic conditions (Hansen, 2011). Good harvests can be expected when rainfall of 300-600 mm is optimally spread throughout the vegetation period. During each of its development stages, the plant is highly susceptible to water-logging, and can therefore only thrive during moderate rainfall, or when irrigation is carefully controlled in drier regions (Naturland, 2002). The water requirement can be met from available soil moisture at sowing, rainfall during the growing season and irrigation (Hansen, 2011). This is because due to its tap roots, the plant is highly resistant to drought, and can provide good harvests even when only stored soil water is available. When irrigated, or during summer rain spells, sesame grows better in sandy than in heavy soils. This is due to its sensitivity to high soil moisture contents (Naturland, 2002).

Sesame needs long periods of sunshine, and is generally a short-day plant – whereby varieties exist which are unaffected by the length of the day (Naturland, 2002). Sesame needs a constant high temperature, the optimum range or growth, blossoms and fruit ripeness is 26-30°C. The minimum temperature for germination is around 12°C, yet even temperatures below 18°C can have a negative effect during germination (Naturland, 2002). Pollination and the formation of capsules is inhibited during heat-wave periods above 40°C. In regions visited by strong, hot winds, the plants only form smaller seeds with a lower oil content. Sesame is sensitive to strong winds when the main stem is fully grown. Tall varieties should not be planted in regions which have strong winds during the harvesting season (and, if necessary, hedges to protect against the wind should be planted) (Naturland, 2002). The response of sesame to both temperature and daylength indicates that it should be cultivated in the wet season in the tropics, or in the summer in the warmer temperate areas. While there is some variation between cultivars, the base temperature for germination is about 16ºC (Bennet, 2011). This warm-season annual crop is primarily adapted to areas with long growing seasons and well-drained soils (Hansen, 2011, www.nutrition-and-you).

Sesame is an annual plant, which grows either bush-like or upright, depending on variety. The plant is usually 60 to 120cm and bears plenty of pink-white color fox-glove type flowers (Figure 3). The pod or fruit which is a dehiscent capsule held close to the stem, appears soon containing white, brown or black seeds depending up on the cultivar type, arranged in rows inside. Each pod (2-5 cm in length) is a long rectangular box like capsule with deep grooves on its sides and may contain up to 100 or more seeds (www.nutrition-and-you). When ripe, the capsule shatters to release a number of small seeds. The seeds are protected by a fibrous 'hull' or skin, which may be whitish to brown or black depending on the variety. 1000 seeds weigh some 4-8g (Chemonics, 2002).

Fig. 3. A Picture of the Sesame Plant with the Seeds

4. Harvesting and post harvesting handling of sesame seed

Sesame seed is harvested when about 50% of capsules turn yellow in colour from green. Other indications of the optimum time for harvesting (physiological ripeness) include; lowest capsules turning brown and beginning to pop open, stem turning yellow, leaves beginning to fall off, end of blossoming, leaves turning yellow (Kimbonguila et al., 2009). Harvesting should not be delayed in order to prevent seed loss through shattering. The size and shape of sesame seed (ie small and flat) makes it difficult to move much air through it in a storage bin. Therefore, the seeds need to be harvested as dry as possible and stored at 6 percent moisture or less (Hansen, 2011, Nigeria's Harvest, 2009; Langham et al., 2008). If the seed is too moist, it can quickly heat up and become rancid. High levels of humidity can cause sesame to take on moisture again and go mouldy; it should therefore only be stored for a short while, or in air-tight containers. If the critical 6% cannot be reached by using only sun drying then artificial methods must be employed. This is because sesame above 5.1% starts getting docked for moisture weight, and above 6.9% moisture, there are quality discounts.

Sesame is generally harvested manually by cutting stems with sticks, and then left to dry for the first 2-3 days after cutting in a windrow (Naturland, 2002). The leaves dry out quickly there, making it easier to bundle them into sheaves. Plants are tied with a rope into small bundles or sheaves (diameter of 15 cm, bottom: 45-80 cm) and positioned erect on a mat or tarpaulin for drying to complete. This prevents seed wastage ns contamination. The sheaves should be positioned so that the sun can shine down directly onto the capsules. This results in less drying time, better heat and air circulation, avoidance of fungi infection, ease of turning when shaking bundles, more extensive shaking/threshing with fewer losses. The sheaves should not need to be dried for longer than 15 days. After the sheaves have dried out fully, they are tipped out onto sturdy cloths or canvases and threshed with sticks. The cloths/canvases should be at least 6 m², to avoid contamination with stones and soil.

Mechanical harvest is better, as the unripe plants are cut, and then the pre-dried sheaves threshed out. This reduces the amount of seeds lost, and the hay makes better fodder. Most threshing machines with a sheaf pick-up function are suited to the task. Varieties that open are easier to thresh mechanically than those that remain closed, as less force is needed. . Sesame seed is easily threshed and relatively delicate, so drum speed should be reduced to about half of that required for cereals, and the concave clearance made as wide as possible. Seed damage during harvesting affects both the viability of the seed, storage and the quality of the oil.

After the seeds have been harvested and dried, the storage sacks must be checked and free of insects. Packaged sesame should be stored in a dark place at low temperatures (below 18°C) and low relative humidity. Also sesame contain more of unsaturated fats hence should be stored in air-tight containers to avoid them turning rancid. Under optimum storage conditions, sesame can be stored for several months even up to 1 year. For safe long-term storage, sesame seed should be clean, have moisture content no more than 6% and be stored at a relative humidity of approximately 50% and at a temperature less than 18°C (Bennet, 2011).

5. Nutritional benefits of sesame seed

Sesame seed (*Sesamum indicum L.*) is an oilseed with a chemical composition of about 50-52% oil, 17-19% protein and 16-18% carbohydrate (Tunde-Akintunde and Akintunde, 2004). Its seed contains about 42-54 % quality oil, 22-25 % protein, 20-25 % carbohydrates and 4-6% ash. The hull contains large quantities of oxalic acid, crude fiber, calcium and other minerals. When the seed is properly dehulled, the oxalic acid content is reduced from about 3 % to less than 0.25 % of the seed weight (Akinoso et al., 2010). Sesame seed contains antioxidants which inhibit the development of rancidity in the oil. In the food industry, where synthetic antioxidants are used extensively, there is an increasing demand for more of these natural products (Bennet, 2011). The nutritional benefits derived from sesame seeds are based on the variety being utilized.

Sesame seed (*Sesamum indicum* L.), from Northern Congo were reported to contain 5.7% moisture, 48.5% crude oil, 20% crude proteins, 7.78% carbohydrate (by difference), 9.4% crude fiber and 4.2% ash (Nzikou et al., 2009; 2010). The protein content of White Sudan sesame seed was high (~25%) similar to other foodstuffs rich in proteins such as almond, hazelnut protein the contents of which were respectively, 20% and 21% (Borchani et al., 2010). The ash content in raw sesame was relatively high (~5%) compared to other products of great consumption such as almond (3%), and the pistachio (2.7%) (Borchani et al., 2010). Other Sudanese local and improved varieties considered by El Khier et al (2008) had protein content of 32.50 to 35.94 and 33.43 to 40.00 respectively.

The seeds also contained significant amount of important minerals with the Potassium concentration being the highest, followed by Phosphorus, Magnesium, Calcium and Sodium (Loumouamou et al., 2010). For White sesame seed (*S. indicum* L.) from Sudan, oil was 52.24%, protein 25.97%, fibre 19.33% and ash 4.685 (El Khier et al, 2008). The predominant mineral composition was calcium followed by potassium, magnesium and phosphorus. All other elements were present in comparatively low concentrations (Elleuch et al., 2007). This is similar to the results obtained by Borchani et al. (2010) for white Sudanese sesame. Potassium is an essential nutrient and has an important role is the synthesis of amino acids and proteins. Calcium and Magnesium plays a significant role in photosynthesis, carbohydrate metabolism, nucleic acids and binding agents of cell walls. Calcium assists in teeth development Magnesium is essential mineral for enzyme activity, like calcium and chloride; magnesium also plays a role in regulating the acid-alkaline balance in the body. Phosphorus is needed for bone growth, kidney function and cell growth. It also plays a role in maintaining the body's acid-alkaline balance. The presence of these minerals also confirms the fat that sesame seed is of high nutritional benefit to its consumers.

6. Description / Physical properties of sesame seed

The seed characteristics or physical properties of sesame seeds vary and this variation may likely be as a result of variability in genotypic effects (El Khier et al., 2008). The physical properties i.e length, width, thickness, geometric mean diameter, sphericity and surface area of the two common local sesame seeds varieties in Nigeria varied from 2..9 – 3.2mm, 1.9 – 2.1mm, 0.85 – 0.91mm, 1.59 – 1.72mm, 0.575 – 0.58 and 7.05 – 10.2mm² respectively (Tunde-Akintunde and Akintunde, 2007). The values obtained for the local varieties were close to that obtained for improved varieties obtained at the National Centre for Genetic Resources and Biotechnology (NACGRAB), Moor plantation, Ibadan, Oyo State, Nigeria which include NCRI BEN 0IM, NGB\04\026, NCRI BEN 03L, NG\SA\07\179, NG\SA\07\090, NG\SA\07\095, NG\SA\07\106, NG\SA\07\052, NG\SA\07\137, OM1, OKENE MKT, NCRI BEN 02M, KANO 05. The length of the improved varieties varied from 2.26mm – 3.01mm, width 1.55 – 1.86mm, thickness 0.74 – 0.97mm and the geometric mean diameter varied from 1.32 – 1.7mm. The variation in degree of sphericity and that of surface area was from 0.57 – 0.64 and 5.84 – 8.94 mm² respectively (Adebowale et al., 2010). The length, width, and thickness of a local sesame seed variety Giza 32 from the Agricultural Engineering Research Institute, Egypt was 2.5, 1.65, and 0.94 mm respectively while the geometric diameter, sphericity and flat surface area were 1.57 mm, 62.84 % and 3.24 mm² respectively (Arafa, 2007).

7. Utilization of sesame seed

Sesame is grown for its seeds, and the primary use of the sesame seed is as a source of oil for cooking. The young leaves may also be eaten in stews, and the dried stems may be burnt as fuel with the ash used for local soap making, but such uses are entirely subordinate to seed production (Table 2). In West Africa, the shoots and young leaves of the varieties *S. alatum* and *S. radiatum* are eaten as a vegetable.

Input	Products	Description and Uses
Seeds	Confectionery	Fried seeds may be bound together with sugar syrup to give sweetmeats.
Seeds	Biscuits	The whole seeds can be baked into biscuits.
(Hulled) seeds	Bakery	Popular in northern Europe either incorporated into breads or as decorative toppings. May be used hulled or whole.
Seeds, sometimes roasted	Oil	Particularly used in oriental cuisine. The flavor is quite strong and rarely compatible with traditional Western style cooking but also used as a salad oil.
Oil	Medicinal treatment	Ulcers and burns
Oil	Margarine	Once an important use, now other cheaper vegetable oils are available
Oil	Aerosol	Reported use as a synergist for pyrethrum sprays
Low grade oil	Various	Soaps paints, lubricants, and illuminants. Local uses, of no importance in international trade
Hulled seeds	Tahini	A paste of sesame seeds which is used as an ingredient in eastern Mediterranean and Middle Eastern foods
Tahini	Dips & spreads	Various ingredients, such as chickpeas or eggplants, are added to tahini to make dips and spreads such as hummus
Tahini	Halva	A sweet made from tahini and sugar with other added flavorings
Cake	Animal feed	Protein rich useful supplement
Cake from hulled seeds	Ingredient	Used in some Indian cooking. Also as a snack in, for example, the Nigerian Kulikuli.

Source: Chemonics International Inc. (2002)

Table 2. Sesame Products

Sesame is commercialized in a number of forms. Most sesame is processed directly into oil by the grower or within the producing region, but can also be sold in various stages of processing, for various uses, such as meal, paste, confections, and bakery products. Sesame seeds can also be consumed directly as a highly nutritious foodstuff (Naturland, 2002). Sesame seeds have delicate nutty flavor. Their flavor indeed becomes more pronounced once they are gently roasted under low flame just for few minutes. De-hulled sesame seed is mainly used to add texture, taste and aesthetic value to a variety of bakery products like bread, bread sticks, cookies, sesame bars etc; and also as an additive to cereal mixes and crackers. It is also used in the making of tahin or sesame butter - a paste of ground sesame seeds, which is used as an ingredient (in Greece) and halva, placed within breads or sprinkled on the surface of bread and breadsticks as a garnish (Germany and the Netherlands) and for the preparation of rolls, crackers, cakes and pastry products in commercial bakeries (Nzikou et al., 2009). Ground and processed seeds can also be used in sweet confections, candies are made from sesame mixed with honey or syrup and roasted (in South Asia, middle East and East Asia) while sesame paste and starch are used to make *goma-dofu* (Japan). Sesame seed can also be in the manufacture of margarine, sprinkled over salads and desserts, particularly sundaes and other ice cream based preparations, preparation of gomshino (a Japanese delicacy) and soybean oil. It can also be used in other food dishes including Mexican and East Asian cuisines.

Sesame seed is primarily grown for its oil in Nigeria and the oil is a primary source of cooking oil in Eastern Nigeria. The major portion of sesame seed produced in countries like Nigeria and India is used for extraction of oil. Sesame oil is mostly used as traditional cooking oil in Chinese food items and in Japan. Sesame seed is an excellent source of high quality oil and protein, its oil is odourless and close in quality to olive oil (Tunde-Akintunde and Akintunde, 2007). Sesame oil has no odour, it is straw-like in colour and has an excellent taste. Sesame seed oil is a natural salad oil, requiring little or no winterization, is one of the few vegetable oils that can be used directly without refining and is used widely as cooking oil. Because of the excellent quality of the edible oil it produces, sesame is often called queen of the oil seed crops. Light sesame oil have a high smoke point and is suitable for deep-frying, while dark sesame oil (from roasted sesame seeds) has a slightly lower smoke point and is unsuitable for deep-frying. Instead it can be used for the stir frying of meats or vegetables, or for the making of an omelette. East Asian cuisines often use roasted sesame oil for seasoning. It is also used widely for production of magarine, shortening, canned sardine and beef as well as in soap and confectionary industries (NAERLS, 2010). Sesame oil has a high preservative effect though the seeds are prone to rancidity because of its high oil content. The oil prevents rancidity due to a preservative within the oil called sesamol. Sesame oil obtained during the first, cold pressing is one of the costliest produced. The oil is light yellow, does not dry out, and can be used with strong heat. Sesame oil obtained from the second, warm pressing and extraction has a lower quality than cold-pressed.

In the industry, sesame oil may be used as a solvent in injected drugs or intravenous drip solutions, a cosmetics carrier oil, to coat stored grains to prevent weevil attacks. The oil also has synergy with some insecticides. Lower grade sesame oil can be used locally in soaps, lubricants, and illuminants. Sesame oil can also be used as a raw material in the manufacture of inks (sesame oil yields a top quality ink after it is burnt), paints, and pharmaceuticals (as healing oil or a vehicle for drug delivery). The oil also has additional use in the industrial preparation of perfumery, cosmetics (skin conditioning agents and moisturizers, hair preparations, bath oils, hand products and make-up), insecticides and paints and varnishes. However, all of these uses are comparatively insignificant in terms of the quantities used.

The seed is typically crushed intact for the oil. This, however, yields a meal that is made bitter and somewhat indigestible by the presence of the fibrous husk. As such the meal is only useful as cattle feed. The quality of the meal can however be improved by removing the seed coat, dehulling, before crushing. In India, where sesame meal is an important food, this process is a standard feature of an oil extraction plant. The meal remaining after oil extraction has unique nutritional properties. It has a high protein concentration which is rich in methionine, cysteine and tryptophan. Since these amino acids are missing from a number of other sources of vegetable protein, such as soya, sesame meal or flour can be added to recipes to give a better nutritional balance to health food products thus complementing most oil seeds and vegetable proteins very well. The seed cake is also an excellent protein supplement in the animal feed industry. The press cake can also be used for food enrichment of infant weaning foods (NAERLS, 2010).

Different cultures have different traditional uses for sesame. In Nigeria, industrial processing and utilization of sesame have not been fully developed. However, the product is locally processed and utilized in various forms in the states where the crop is cultivated. Principal among the products are: *"Kantun Ridi"* and *"Kunun Ridi"*. At the local level, oil is also extracted from the seed and the cake is made into *"Kulikuli"* which together with the leaves are used to prepare local soup known as *"Miyar Taushe"*. The oil is used locally for cooking as well as for medicinal purposes such as the treatment of ulcers and burns. The stem and the oil extract are equally used in making local soap. In all the states where sesame is cultivated, women are more involved in the local processing of sesame seed into commercial products.

In Japan, tan and black sesame seed varieties are roasted and used for making the flavoring gomashio. In Greece the seeds are used in cakes, and in Togo they are a main soup ingredient. The seeds are also eaten on bread in Sicily and France (called *"ficelle sésame"*, sesame thread). In Congo and Northern Angola, ground sesame or *wangila* is a delicious dish, especially when cooked with smoked fish or lobsters. In Manipur (North Eastern State of India), black sesame is used in the preparation of a favorite side dish called 'Thoiding' and in 'Singju' (a kind of salad). Also in Tamil Nadu, India, a ground powder, *Milagai Podi*, made of sesame and dry chili is used to enhance flavor and consumed along with other traditional foods

8. Oil characterization

The different variety of sesame plant, cultivation climate, soil type, ripening stage, the harvesting time of the seeds and the extraction method used has an important effect on the yield and quality of oil obtained from sesame (Nigeria's Harvest, 2009; El Khier et al., 2008). Black seed types have been reported to contain less oil than white and brown seeded strains (Akinoso et al., 2010). Composition of fatty acids in sesame oil was also reported to be variable among the different cultivars worldwide (El Khier et al., 2008).

The chemical properties of oil are amongst the most important properties that determines the present condition of the oil. Free fatty acid and peroxide values are valuable measures of oil quality. The chemical properties of white and black sesame seed varieties from Hong, Adamawa State, Nigeria was determined. The unsaponifiable matter (% of oil), saponification value (mg KOH/g), iodine value (g/100 g) and free fatty acid (% Oleic) was 0.76, 150.26, 98.20 and 0.78 while that of the black variety was 0.71, 158.04, 106.26 and 0.73 respectively (Fariku et al., 2007). Oil obtained from another Nigerian variety (Goza-25) had the following properties; peroxide value varied from 3.9 and 15.4 meq/kg while oxidative stability varied from 13 h to 63.3 h (Akinoso et al., 2010). These values are significantly different from values obtained from varieties in other locations.

The iodine value of some local Sudanese and imported sesame seed cultivars varied from 101.52 to 114.85g/100g for the local cultivars and 97.70 to 111.30 g/100g for the introduced cultivars (El Kheir et al., 2008). Saponification value varied from 174 – 196.32 mg/g for local and 182.31 - 198.02mg/g for improved, while the Peroxide value varied from 2.22 – 15.07 meq/H_2O_2/Kg for local and 2.24 – 10 meq/H_2O_2/Kg for improved. The acid value however varied form 3.1 – 6.6 mg/g for local and 3.1 – 9.3mg/g for introduced. The chemical properties of oil obtained from white Sudanese sesame seed are iodine value 113.35 g of I_2 100 g^{-1} of oil, saponification value 186.60 mg KOH g^{-1} oil, peroxide value 0.14 meq O_2 kg^{-1} oil, acid value 1.64 mg KOH g^{-1} oil, and free fatty acids 0.82 % of Oleic acid (Borchani et al., 2010). Oil obtained from solvent extraction of Northern Congo sesame seed had the following chemical properties; iodine value of 117.2 w ijs, free fatty acid value of 0.06 % oleic acid, saponification value of 197 and peroxide value of 0.06 (Nzikou et al., 2009; 2010).

Generally the high iodine values of sesame oil is an indication of the presence of unsaturated fatty acid and this places the oil in the drying groups, The low value of free fatty acid shows that this oil is stable. The higher oxidative stability of raw sesame oil could be attributed to the presence of such natural antioxidants as tocopherols, sesamin and sesamolin (Elleuch et al., 2007). The saponification value is high and this suggests the use of the oil in production of liquid soap, shampoos and lather shaving creams. the peroxide value is lower than that expected of rancid oil, which ranges from 20.00 to 40.00 mg.gG^1 oil (Nzioku et al, 2010). This shows that sesame oil is not rancid is considered stable.

9. Processing methods of sesame seed and products

Sesame seed processing is basically done to clean and dehull seed as well as to extract oil from seed. Sesame can be processed to several different stages, such as simply cleaning, or cleaning and dehulling, cleaning/dehulling/drying, cleaning/dehulling/drying/crushing for oil, etc. Generally, sesame seeds are cleaned, dehulled (important because of presence of tannins which are located in hulls). In Nigeria, dehulling is done by soaking in a salt solution overnight. Seeds are rubbed in a mortar to loosen pericarp and then kernel is separated from oat by sedimentation washing (NAERLS, 2010). After dehulling, seeds are washed and then dried usually with sun-drying.

Conditioning of oil seeds is an important operation in the production line of sesame oil. These activities include roasting, flaking, size reduction, cooking, pre-pressing and drying. Dehydrating and roasting of sesame seeds before oil expression improves sesame oil yield and quality. Also the oxidative stability of oil and by-products has been reported to depend on processing techniques and variety of seeds (Akinoso et al., 2010). Beniseed oil extraction is done traditionally in Nigeria by pounding the seeds in a mortar and pouring water into it (Tunde-Akintunde and Akintunde, 2007). The oil floats to the surface from where it can be removed by skimming. This method is slow and laborious and results in low oil yield. Other traditional methods involve crushing to paste using a local grinding machine. Boiling water is added to the paste, stirred and left for 24h. The oil floating on top of the paste is decanted and the process is repeated until negligible oil is formed (Fariku et al., 2007). Another oil extraction method is to roast seeds for 5 – 19 minutes at 180 – 210ºC and then mill. Oil is pressed out by adding water to the milled product (NAERLS, 2010). After oil extraction, the cake is dried by sun drying and milled to obtain defatted flour.

The processing of sesame products in the US is similar but the facilities used are different (Hansen, 2011). After harvesting, the seeds are cleaned and hulled. The seeds pass through an air separation stage to remove any foreign particles. About 10 percent of this "cleaned natural seed" moves directly into food use as whole seed to be blended into flour for baked goods. Next, a combination of water and friction work together as the seeds are passed against the chamber of the hulling machine to separate the hull from the seeds. This dust-free de-hulled seed makes up 30 percent of domestic production and has a 99.97 percent purity for the baked goods market. Once the seeds have been hulled, they are passed through an electronic color-sorting machine that rejects any discolored seeds to ensure perfectly colored sesame seeds. Immature or off-sized seed is removed but saved for oil production. Sesame oil is extracted by pressure in a mechanical expeller and is tolerant of only minimal heating by the extraction process. This pure, mechanically expressed oil is called "virgin" oil and is preferred by many food handlers. The oil is often blended with other vegetable oils for salads and other food uses. Sesame oil should be kept refrigerated. Sesame seeds can become rancid if exposed to prolonged heat. If properly stored, the packed seeds have a 2-year shelf life with little reduction in quality (Hansen, 2011). A dehulling method used for sesame in India is usually done by soaking the seeds overnight in water, followed by drying and rubbing against a rough surface. The separated hulls are removed by winnowing. This method is also laborious, time consuming and suitable for processing small quantities only. A more convenient dehulling technique has been developed through addition of 3% Sodium Chloride (salt) and soaking over night (Chemonics, 2002).

The dehulled seed can be expeller pressed for obtaining good quality oil. The cake is further subjected to solvent extraction to recover the residual oil and the protein rich cake is used for protein fortification of various food preparations. An oil extraction process in India involves preliminary cleaning and grading, placing in a boiling solution of sodium hydroxide for a prescribed time and then thoroughly washing by a stream of water. The hulls are removed by washing and brushing seeds under a current of water. The dehulled wet seeds are then dried in a cross-flow or fluidized bed drier (www.nisc).

Sesame oil can be extracted from sesame by solvent extraction or mechanical expression. A solvent extraction procedure (soxhlet method) was reported by Nzioku et al. (2010). Dried sesame seeds w ere ground in the dry mill of a blender. 50 g of ground seeds were placed into a cellulose paper cone and extracted using light petroleum ether (b.p 40-60°C) in a Soxhlet extractor for 8 h. The oil was then recovered by evaporating of the solvent using rotary evaporator and residual solvent was removed by drying in an oven at 60°C for 1 h and flushing with 99.9% nitrogen. The hot pressed oils are usually refined before consumption to remove free fatty acids, residues and all aromatic compounds resulting in a bland colorless oil. Refined oils are suited to the cooking of the Western hemisphere where highly aromatic oils are not appreciated.

10. Derivable benefits for sesame seed processing

Sesame seeds are widely considered to be healthful foods. They are high in energy but contain many health benefiting nutrients, minerals, antioxidants and vitamins that are essential for wellness and have positive effects on human health (www.nutrition-and-you, Borchani et al. 2010). They are very good sources of B-complex vitamins such as niacin, folic acid, thiamin (vitamin B1), pyridoxine (vitamin B6) and riboflavin. The seeds are incredibly rich sources of many essential minerals which include Calcium, iron, manganese, zinc, magnesium, selenium and copper in concentrated amounts. Many of these minerals have vital role in bone mineralization, red blood cell production, enzyme synthesis, hormone production as well as regulation of cardiac and skeletal muscle activities (www.nutrition-and-you). The seeds are especially rich in mono-unsaturated fatty acid *oleic acid* which comprise up to 50% fatty acids in them. Oleic acid helps to lower LDL or "bad cholesterol" and increase HDL or "good cholesterol" in the blood. The seeds are also very good source of dietary proteins with fine quality amino acids that are essential for growth, especially in children. In addition, sesame seeds contain many health benefiting compounds such as sesamol and sesaminol which are phenolic anti-oxidants and help stave off harmful free radicals from the body. As a result of the benefits of sesame, seed exportation offer highest potential for farmers due to the demand for the seed, but Sesame's potential for commercial processing in Nigeria is also great (Nigeria's Harvest, 2009). This is because sesame can be processed into various forms which include oil, meal, paste, confections, and bakery products.

These products of sesame have a number of benefits that can be derived from processing the seeds. The nutritional value of the meal makes it a potential source of livestock feed, this is because it is a relatively good source of CP- crude protein (Omar, 2002). Studies carried out on incorporation of sesame oil cake in rations had positive effects on calves' performance (Omar, 2002). Thus the use of sesame cake in areas where sesame is produced will be beneficial to farmers. The sesame paste contains high monounsaturated to saturated fatty acid ratios, and might be desirable substitutes for highly monounsaturated oils such as olive oil in diets (Borchani et al., 2010). Thus the paste can be utilized as a potential source of edible oils for human consumption.

Raw sesame oil has a high free fatty acids content and is stable in oxidative conditions. The high percentage of oil in the seed makes this seed a distinct potential for the oil industry (Nzioku et al., 2010). This oil may find application as a raw material in industries for the manufacture of vegetable oil-based ice cream (Ibiyemi et al., 1992). All these characteristics lead to more diverse and novel applications of sesame oil in food, cosmetics and pharmaceutical products. Apart from these applications, raw oil from sesame also has a lot of potential as a renewable resource considering its viscosity which is close to those of soybean and sunflower (Fariku et al., 2007). Its flash point is lower but comparable to values reported for soybean and sunflower oils. The calculated fuel value is also comparable to those of soybean and sunflower oils hence, it has high energy density. The iodine value obtained for sesame seed oil also shows that it is a non drying oil and as such it is unsaturated thus making it suitable for utilization as fuel as well as raw material in industries for the manufacture of soap and vegetable oil – based ice cream (Fariku et al., 2007).

Sesame oil has a high percentage of polyunsaturated fatty acids (omega-6-fay acids) but keeps at room temperature uniquely due to the presence of sesamol and sesamin, two naturally-occurring preservatives. The presence of high polyunsaturated fatty acids content make it possible to use sesame oil for cooking in place of other edible oils and to help reduce high blood pressure and lower the amount of medication needed to control hypertension. Sesame oil can also be used for medicinal purposes which include reduction of cholesterol levels, anti-bacterial effects, and even slowing down certain types of cancer (www.nutrition-and-you). Apart from these various constituents present in the sesame oil have anti-oxidant and anti-depressant properties.

11. Adverse effects

Though sesame seeds have a wide range of health and commercial benefits, they have some anti-nutritional properties. Sesame seeds contain a high amount of the phytic acid which is an anti-nutrient. Another disadvantage of the seed is that it produces allergic reactions in some people. The allergy may be mild and appear as hives, dermatitis and itching or be severe and lead to severe physical symptoms like vomiting, pain abdomen, swelling of lips and throat leading to breathing difficulty, chest congestion and death. The laxative effect of sesame also indicates that sesame oil should not be used by people who have diarrhea.

12. Further studies

Based on different research studies carried out on sesame seed, the following studies can be further carried out on sesame seeds and its products to enhance its utilization especially in Nigeria:

- The seed could be further explored to develop a high crude oil yield variety that would be of immense nutritional and economic advantage (Adebowale et al, 2010)
- The high percentage of oil makes this seed a distinct potential for the oil industry (Nzikou et al., 2009).
- The high unsaponifiable matters content (1.87%) of sesame oil guarantees the use of the oil in cosmetics industry. The oil extracts exhibited good physicochemical properties and could be useful for industrial applications (Nzikou et al., 2010).
- Extensive research work still needs to be carried out on sesame seed for its industrial utilization as a biofuel resource material in Nigeria (Fariku et al., 2007)

13. Conclusion

Sesame seed is an oil-seed crop, with edible and odourless oil, and with good source of protein for man and livestock. Its utilization in Nigeria is however mainly for exports and limited local household use within regions in which it is grown in the nation. This is because the low level of information on the Nigerian variety has limited its utilization nationally but information available on international varieties has however made its utilization as an export crop more prominent than being used for home consumption.

This chapter is aimed at documenting relevant information of Nigerian variety of sesame seed (making a comparison with international varieties) and also examining and presenting the results of studies that have been carried out on the oilseed generally. The information thus documented is expected to bridge the gap and provide people involved in the food chain of seame seed- with useful literature on the crop. This information will also enhance utilization of the crop and enable producers, processors and consumers to effectively exploit the potential benefits of sesame seed with regards to the Nigerian food and agricultural system.

14. References

Adebowale, A.A, Sanni S.A., & Falore O.A. (2010). Varietal differences in the Physical properties and proximate composition of elite sesame seeds. *Libyan Agricultural Research Center Journal International* 1 (2): 103 – 107

Akinoso, R, Aboaba, S A & Olayanju, T.M.A. (2010) Effects of Moisture Content and Heat Treatment on Peroxide Value and Oxidative Stability of Un-Refined Sesame Oil. *AJFAND* 10 (10): 4268- 42850

Arafa, G. K. (2007). Some Physical and Mechanical Properties of Sesame Seeds Concerning the Selection of Separtion Unit. *Misr Journal of Agricultural Engineering*, 24(2): 415-429

Bennet, M. (2011). Sesame seed: A Handbook for Farmers and Investors. 29/08/11. Available at www.agmrc.org/media/cm/sesame_38F4324EE52CB.pdf

Borchani C., Besbes S., Blecker C. H. & Attia H. (2010). Chemical Characteristics and Oxidative Stability of Sesame Seed, Sesame Paste, and Olive Oils. *Journal of Agriculture, Science and. Technology*. 12: 585-596

CBN (2009). Annual Report and Statement of Account for the Year Ended 31st December 2008. *Central Bank of Nigeria*, 2009; 142-145.

Chemonics International Inc. (2002). *Overview of the Nigerian Sesame Industry*. The United States Agency for International Development (USAID)/Nigeria RAISE IQC Contract No. PCE-I-00-99-00003-00

Dehulled sesame seed (n.d.). Publication of National Small scale Industries corporation. 20/07/11. Available at http://www.nsic.co.in/schemes/documents/projprofiles/DEHULLED%20SESAME%20SEED.pdf

El Khier, M. K. S., Ishag K.E.A., & Yagoub A. E.A. (2008). Chemical Composition and Oil Characteristics of Sesame Seed Cultivars Grown in Sudan. *Research Journal of Agriculture and Biological Sciences*, 4(6): 761-766.

Elleuch, M., Besbes, S., Roiseux, O., Blecker, C. & Attia, H. (2007). Quality characteristics of sesame seeds and by-products. *Food Chemistry* 103 (2): 641 – 650

Fariku, S., Ndonya, A.E., & Bitrus, P.Y. (2007) Biofuel characteristics of beniseed (*Sesanum indicum*) Oil. *African Journal of Biotechnology* Vol. 6 (21), pp. 2442-2443

Gandhi A.P. (2009) Simplified process for the production of sesame seed (*Sesamum indicum* L) butter and its nutritional profile. *Asian Journal of Food and Agro-Industry*. 2(01), 24-27

Global AgriSystems. (2010) Dehulled and roasted sesame seed oil processing unit. 18/08/11 Available at http://mpstateagro.nic.in/Project%20Reports%20pdf/Dehulled%20and%20Roasted%20S esame%20Seed%20Oil%20Processing%20Unit.pdf

Hansen, 2011, R. (2011) Sesame Profile. 19/08/11. Available at http://www.agmrc.org/commodities__products/grains__oilseeds/sesame_profile.cfm

Ibiyemi, S.A., Adepoju, T.O., Okonlawun, S.O., & Fadipe, V.O. (1992). Toasted *Cyperus esculentus* (tigernut): Emulsion preparation and stability studies. *Nigerian Journal of Nutrional. Sciences* 13(1-2): 31-34.

Jospeh, H. (2009) Nigeria: Boosting Benue's Beniseed Potential. Daily Trust newspaper. 30 July 2009

Kimbonguila, A., Silou, T.H., Linder, M. & Desobry S. (2009) Chemical Composition on the Seeds and Oil of Sesame (*Sesamum indicum* L.) Grown in Congo-Brazzaville. *Advance Journal of Food Science and Technology* 1(1): 6-11.

Langham D. R, Riney J., Smith G., & Wiemers T. (2008) Sesame Harvest Guide. September 2008. 21/08/11. Available at
    www.sesaco.net/Harvest%20pamphlet%20080903%20final%20b.pdf
Lee, J.Y., Lee, Y.S., & Choe, E.O. (2008) Effects of Sesamol, Sesamin and Sesamolin Extracted from Roasted Sesame Oil on the Thermal Oxidation of Methyllinoleate. *Food Science and Technology* 42: 1871-1875.
Loumouamou B., Silou T.H. & Desobry S. (2010). Characterization of Seeds and Oil of Sesame (*Sesamum indicum* L.) and the Kinetics of Degradation of the Oil During Heating. *Research Journal of Applied Sciences, Engineering and Technology*, 2(3): 227-232.
NAERLS (2010). Beniseed production and utilisation in Nigeria. Extension Bulletin No 154, Horticulture Series No 5. 17/07/11. Available at
    www.naerls.gov.ng/extmat/bulletins/Beniseed.pdf
Naturland. (2002) Organic farming in the Tropics and Subtropics: Sesame. 21/07/11. Available at
    www.naturland.de/fileadmin/MDB/documents/Publication/English/sesame.pdf
Nigeria's harvest (2009). Small seed provides large income. Volume 5, October 21, 2009 http://www.nigeriamarkets.org/files/Nigeria%27s_Harvest_Vol_5_Exporting_Nigerian_Sesame.pdf
Nzikou, J.M., Mato,s L., Bouanga-Kalou, G., Ndangui, C.B., Pambou-Tobi, N.P.G., Kimbonguila, A, Silou Th., Linder, M. & Desobry, S. (2009) Chemical Composition on the Seeds and Oil of Sesame (*Sesamum indicum* L.) Grown in Congo-Brazzaville. *Advance Journal of Food Science and Technology* 1(1): 6-11.
Nzikou, J.M., Mvoula-tsiéri, M., Ndangui, C.B., Pambou-Tobi, N.P.G., Kimbonguila, A., Loumouamou, B., Silou, Th. & Desobry, S. (2010) Characterization of Seeds and Oil of Sesame (*Sesamum indicum* L.) and the Kinetics of Degradation of the Oil During Heating. *Research Journal of Applied Sciences, Engineering and Technology*, 2(3): 227-232.
Omar J. M. A. (2002). Effects of feeding different levels of sesame oil cake on performance and digestibility of Awassi lambs. Small ruminant research. 46(2), 187 – 190
Overview of the Nigerian Sesame Industry. 2002.
    www.nigeriamarkets.org/files/sesame_subsector_overview.pdf
Sesame oil. http://en.wikipedia.org/wiki/Sesame_oil
Sesame seeds nutrition facts. www.nutrition-and-you.com/sesame-seeds.html
Sesame. http://en.wikipedia.org/wiki/Sesame
Tunde-Akintunde, T.Y., & Akintunde, B.O. (2004). Some physical properties of Sesame seed. *Biosystems Engineering* 88 (1), 127 – 129
Tunde-Akintunde, T.Y., & Akintunde, B.O. (2007). Effect of Moisture Content and Variety on Selected Properties of Beniseed. *Agricultural Engineering International: the CIGR Ejournal*. Manuscript FP 07 021. Vol. IX.
Weiss EA (2000). Oil Seed Crop. 2nd Edition Blackwell Longman Group Ltd. USA.

3

Oil Presses

Anna Leticia M. Turtelli Pighinelli and Rossano Gambetta
Embrapa Agroenergy,
Brazil

1. Introduction

For human nutrition, vegetable oils and animal fats play an important role, acting as an energy source and supplying the human body with more than twice the calories per unit weight than those provided by proteins and carbohydrates. Other benefits of fats are that they are suppliers of essential fatty acids, which are not synthesized by the human body, but are of great importance for our organism. Vegetable oils act as carriers of group of vitamins (A, B, E and K), help the body to absorb other vital elements from food and are also used to give more flavors to food (Bachmann, 2004).

Regarding oilseed materials, they can be divided into those for the production of edible vegetable oils and protein such as soy, sunflower, canola, palm and olive; those where the oil is a byproduct of fiber production, i.e. cottons; crops for food purposes which also produce oil, like corn, coconut, peanuts and nuts; crops which produce non-edible oils such as castor and Jatropha and finally, sources as microbial products, like algae, that can produce oil (Walkelyn & Wan, 2006).

Although the main use of vegetable oils and animal fats is for human consumption, recently there has been an increased interest in vegetable oils due to its use as feedstock to produce biodiesel, a renewable and less polluting fuel when compared to diesel of fossil origin. Other applications comprise its use as animal feed due to their high protein meal, in medicinal purposes, as lubricant, fuel for lamps and wood preservatives (World Bank Group, 1998).

According to Gunstone (2005), the main components of crude vegetable oils are triacylglycerols, corresponding to approximately 95% of its composition along with some free acids, monoacylglycerols, and diacylglycerols. They also contain variable amounts of other components such as phospholipids, free and esterified sterols, triterpene alcohols, tocopherols and tocotrienols, carotenes, chlorophylls and other coloring matters, and hydrocarbons as well as traces of metals, oxidation products, undesirable flavors, and so on. An important classification of vegetable oils is related to its fatty acid composition. Table 1 shows some vegetable oils with their fatty acid composition. Depending on the concentration of fatty acids present in vegetable oils, they can be classified as follows: lauric oils, palmitic oils, oleic/linoleic oils, high oleic oils, linolenic oils and erucic oils.

An interesting article reviews several important catalytic functionalisations, i.e. heterogeneous and homogeneous catalysis, like additions, reductions, oxidations and metathesis reactions of fatty compounds and glycerol resulting in new attractive products,

Oil source	C16:0 (Palmitic)	C18:0 (Stearic)	C18:1 (Oleic)	C18:2 (Linoleic)	C18:3 (Linolenic)
Cocoa butter	26	34	35	-	-
Corn	13	3	31	52	1
Cottonseed	27	2	18	51	Tr
Groundnut	13	3	38	41	Tr
Linseed	6	3	17	14	60
Olive	10	2	78	7	1
Palm	44	4	39	11	Tr
Canola	4	2	56	26	10
Rice bran oil	20	2	42	32	-
Safflower	7	3	14	75	-
Safflower (high oleic)	6	2	74	16	-
Sesame	9	6	41	43	-
Soybean	11	4	22	53	8
Sunflower	6	5	20	60	Tr
Sunflower (high oleic)	4	5	81	8	Tr

Table 1. Fatty acid composition (Gunstone, 2005)

besides biodiesel. Those products have emerging properties, so they could find a rapid introduction into the chemical market. The fatty acid (methyl or ethyl esters) and glycerol can be directly used after separation units as raw materials. Furthermore, they can build a new basis for new, valuable and sustainable bulk and fine chemicals (Figure 1) (Behr & Gomes, 2010).

In the book "Oilseed crops", Weiss (1983) notes that the history of the processes and methods of extracting oil from oilseeds is fascinating. The earliest record of the oilseed processing is attributed to the Assyrians in 2000 BC, who listed the components of a press to extract oil from sesame seeds. Another interesting historical fact is related to oil milling invention that is attributed to the Apollo´s sons, according to Pliny, who describes in detail the methods employed by his Roman contemporaries to obtain olive oil.

There are two main types of processes for obtaining oil: physical and chemical. The physical process, or expression, involves the use of mechanical power to remove oil from the seed, such as batch hydraulic pressing and continuous mechanical pressing (screw presses). Chemical processes, or extraction, are based in solvent extraction. These processes can be combined in commercial operation, i.e. continuous mechanical pressing (expelling) with continuous solvent extraction, and batch hydraulic pressing followed by solvent extraction (Walkelyn & Wan, 2005; Weiss, 1983). New technologies are emerging, related to the production of vegetable oils, such as supercritical-fluid extraction (Pradhan et al., 2010). "Expression" means the process of mechanically pressing liquid out of liquid-containing solids and "extraction" is the process where a liquid is separated from a liquid-solid system with the use of a solvent (Khan & Hanna, 1983).

The main idea of this chapter is to focus on the processing of oil by continuous mechanical pressing (screw presses), which is a technology widely used today by small oil producers. Though fats can be derived from both vegetable oils and animal fats, in this chapter only

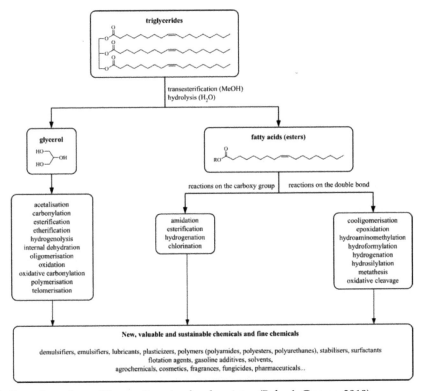

Fig. 1. Overview of possible reactions in oleochemistry (Behr & Gomes, 2010).

vegetable oil will be discussed, due to its healthier aspects when compared to animal fats, leading to an increase in consumption of plants-derived products.

This chapter is organized in three parts:

1. Processes for obtaining vegetable oils
2. Detailing the continuous mechanical pressing; and
3. Examples on the application of pressing for obtaining oil from cotton, peanut and sunflower.

In the first part of this chapter, the commercial methods for the extraction of vegetable oils are reviewed, including: batch hydraulic pressing, solvent extraction and continuous mechanical pressing (also known as screw press). A brief explanation will be given for the new method of producing oil: supercritical extraction. In the second part, the continuous mechanical pressing is detailed. A complete picture of continuous mechanical pressing will be presented, showing how this technology can be optimized for attaining larger oil volumes, having in sight, mainly, small producers of vegetable oils. In the last part of the chapter, some experimental results related to the pressing of cotton, peanut and sunflower seeds are presented. Experiments were optimized in order to obtain a higher volume of oil. The variables studied were speed of rotation of screw press axis, and the best conditions of the raw material, such as moisture content and temperature (Pighinelli, 2010).

2. Processes for obtaining vegetable oils

Commercially, there are three methods used for the production of vegetable oils: batch hydraulic pressing, continuous mechanical pressing and solvent extraction. Each of these technologies will be detailed below

The first step of the process, common to all technologies, is the preparation of raw material. The main unit operations used are shown in Figure 2: scaling, cleaning, dehulling (or decorticating), cracking, drying, conditioning (or cooking), and flaking (Anderson, 2005). Although cooking operation is not included in the flowchart, it is a very important step in the processing of oilseeds. The flowchart below may be altered depending on raw material to be processed.

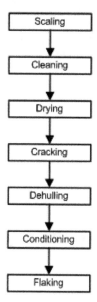

Fig. 2. Unit operations for raw materials preparation (adapted from Anderson, 2005).

Initially the oilseeds are weighed and sent to the cleaning step. A good quality oilseed has around 2% of impurities when it comes from the field. These foreign materials are removed when the oilseeds reach the storage unit and also before starting the extraction of oil. Some examples of foreign materials are a combination of weed seeds, sticks, pods, dust, soil, sand, stones, and tramp metal. Tramp metal is considered hazardous to storage facilities and also for the oil processing operations, so it is the first impurity to be removed, using a magnetic force that pulls these metals from the mass of grain. Sticks and pods are larger and lighter than the oilseeds and can easily be removed by airflow equipment. Weed seeds, sand, and soil are smaller than the oleaginous materials, so those materials can be removed by screening equipment. In the case of peanuts, there are large amounts of stones in grains, and these foreign materials have similar geometry, being necessary to use a gravity separation system to remove impurities (Kemper, 2005).

Due to the high oil content in its composition, oilseeds must have a low moisture content in order to prevent deterioration during storage and also to ensure that downstream unit operations are efficient. This operation is conducted in a dryer (Kemper, 2005).

The main goal of size reduction operation is to increase surface area to facilitate oil removal from the seed inside. This operation must be conducted at proper moisture content. If the moisture level is too low, the seeds are "conditioned" with water or steam to raise the moisture to about 11%. A solvent extraction operation is going to have a higher yield if the flakes are about 0.203 to 0.254 mm. Thinner flakes tend to disintegrate during the solvent extraction process and reduce the miscella percolation rate (Wakelyn & Wan, 2006). Cracking mill is the equipment used to crack oilseeds. This equipment consists of two sets of cylindrical corrugated rolls in series, operating at differential speeds to assist in breaking the oleaginous materials apart (Kemper, 2005).

Another unit operation is dehulling because some oilseeds present outer seed coat known as hull, rich in fiber and poor in oil and protein. The removal of these hulls will produce a better cake with high protein content by weight (Kemper, 2005). Another problem observed with hulled seeds is that the hull will reduce the total yield of oil by absorbing and retaining oil in the cake (Wakelyn & Wan, 2006). Dehulling process removes the lighter hull fraction by aspiration and also the fines, separated from the hulls through various means of hull agitation and screening. A dehulling process can be considered effective if the levels of residual fiber content (from hulls) and residual oil content in the meal fraction were low (Kemper, 2005).

Solvent extraction demands a flaking operation that distorts the cellular structure of the seed and facilitates the percolation of the solvent in the oleaginous material cells. The equipment that performed flaking operation is named flaking mill. Two large diameter rolls in parallel, turning in opposing direction at approximately 250 to 300 rpm., and forced together by hydraulic cylinders. As the oilseeds are pulled through, they are stretched and flattened, forming flakes, in the range of 0.3 to 0.4 mm thick and 8 to 18 mm in diameter (Kemper, 2005).

Cooking is performed prior to extraction and has the following purposes: (1) break down cell walls to allow the oil to escape; (2) reduce oil viscosity; (3) control moisture content (to about 7% for expanding operation); (4) coagulate protein; (5) inactivate enzymes and kill microorganisms; and (6) fix certain phosphatides in the cake, which helps to minimize subsequent refining losses. Cooking temperatures are around 87.8°C during 120 min. An excess in the cooking time will result in a meal with lower nutritional value and can darken both the oil and meal (Wakelyn & Wan, 2006).

2.1 Batch hydraulic pressing

In the late nineteenth century, oilseeds were processed in manual presses, where layers of grains were placed into the equipment, separated by filter cloths and filter press plates and force was applied via a hydraulic cylinder. When the oil has stopped flowing, the workers opened the machine, removed the mass of crushed grain and put more fresh material. At the beginning of the twentieth century the vegetable oil industry worked basically with the hydraulic presses but even making use of a hydraulic cylinder, work with this type of equipment was considered labor intensive. With the emergence of the continuous screw-press, the only application that still requires the hydraulic press is the one that requires gentle handling, such as processing and production of cocoa butter. Presses from this category can be divided into two main groups: open and closed type (Williams, 2005).

Open-type presses: the fresh material, previously prepared and wrapped in press-cloths, is placed between the plates that should be corrugated to assist drainage and overcome cake creepage. In a standard process, it takes 2 minutes to feed the press, 6 minutes to reach maximum pressure, 20 minutes to drain and 2 minutes to remove solids, a total of 30 minutes per batch.

Closed-type presses: in this type, the oilseed is enclosed by a strong perforated steel cage that can apply much more pressure than an open press. Removing oil from the interior of the grain is attained by the pressure applied by a piston placed close to the cage and hydraulically operated. Oil flows through channels that increase in size from inside to outside the cage, thus avoiding any clogging with solid particles (Williams, 2005).

In a typical hydraulic pressing there are three stages as can be identified in Figure 3 (Mrema and McNulty, 1985 in Owolarafe et al., 2008). Initially the loading stage happens before the oil begins to leave the mass of grain (oil point). The application of compressive load causes the seeds to force the air out of the macro pores. This process continues until a critical point that occurs when the seeds respond to pressure through their points of contact. This causes the change in volume and starts the output of oil (initial stage). When the first drop of oil leaks out of the mass, it begins the second stage (dynamic stage), where the air is displaced by the liquid and an air/fluid mixture is extracted. The oil flow increases rapidly to its maximum, which is when the second stage ends. The last stage (final stage) begins when the maximum instantaneous flow rate, i.e. the volume is completely filled with fluid, is reached. Z_1, Z_2 and Z_3 indicate the height of grain layer inside the extractor; P_T is the applied force by the extractor, t_0 is the initial time, taken when the oil starts flowing and T is final time, indicating that the process ends.

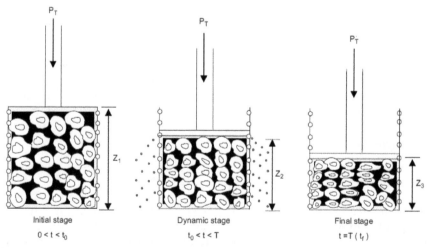

Fig. 3. Stages of hydraulic expressions (Mrema and McNulty, 1985 in Owolarafe et al., 2008).

In 1942 an article entitled "Expression of vegetable oils" was published where a concern in the scientific study and optimization of the hydraulic pressing of oilseeds was expressed (Koo, 1942). For the study, the author used soybean, cottonseed, canola, peanut, tung nut, sesame seed and castor bean. The author evaluated the influence of pressure, temperature and moisture content in oil yield. As the final result it was possible to define an empirical model (on the dry basis) that describes the process (Equation 1).

$$W = CW_o \frac{\sqrt{P\sqrt[6]{\theta}}}{\sqrt{\vartheta^z}} \tag{1}$$

Where: W is the oil yield; C is a constant for one kind of oilseed; W_o is the oil content of the seed; P is the pressure applied; θ is the pressing time; υ is the kinematic viscosity of the oil at press temperature; and z is the exponent on viscosity factor varying from 1/6 to 1/2. Press efficiency can be calculated from the relation W/W_0.

In 2008 an article was published which focused on determination and modeling of yield and pressing rates for the hydraulic press type (Willems et al., 2008). The authors evaluated the influence of pressure profile, temperature, cake thickness and moisture content on the oil yields and rate of pressing for a variety of seeds (sesame, linseed, palm kernel, jatropha and rapeseed). The results showed that when the pressure is increased using a temperature of 100 °C and using the optimum moisture content (close to 2% dry basis), the oil yield increases for all tested oilseeds. The oil yield obtained for the dehulled seeds were considerably higher than that obtained for the hulled seeds. This can be explained by the absorption of oil by fibers present in the hulls. As a final conclusion, the authors stated that when using a press capable of applying a higher pressure (> 45 MPa), the oil yield is increased by 15% (oil / oil) compared with conventional presses.

2.2 Continuous mechanical pressing (screw press)

Among the physical processes for the extraction of vegetable oils, the continuous mechanical pressing emerges as the best technology to serve small farmers. That´s because this type of equipment associates both small scale and low cost when compared to the other methods cited. Another important advantage is the possibility of using cake resulting from the pressing as fertilizer or animal feed, since it is free of toxic solvents.

The operating principle of this equipment consists a helical screw which moves the material, compressing it, and at the same time, eliminating the oil and producing the cake. Optimization of the continuous mechanical pressing consist of defining the optimum parameters, such as temperature and moisture content of grain, or adjustments on the press, in order to reach optimum yields of oil, using a minimum value of pressure applied by the press.

Figure 4 shows the flowchart for the processing of oil by mechanical pressing. More details about continuous mechanical pressing are presented in the next section. Oilseed preparation has already been presented in section 2.

2.3 Solvent extraction

Extraction with solvents is the most effective method for the recovery of oil (almost 98%), especially with materials with low oil content, like soybeans. This method is not indicated to

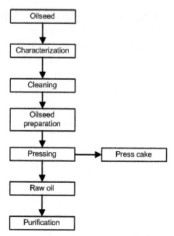

Fig. 4. Screw press scheme (adapted from Jariene et al., 2008).

oilseeds with high oil content, like peanut and sunflower, requiring a prior step of pressing of the seeds and then the cake produced is extracted with a solvent. When performed at low temperature, the solvent extraction has another advantage over screw-pressing: better quality of oil produced. This is because during expelling a sudden heating of the oil can occur, changing some parameters of its quality (Williams, 2005).

One of the disadvantages of the extraction is that the solvent extracts some nontriglycerides, which does not occur in the expelling (Williams, 2005). Another serious problem is the presence of volatile organic impurities in the final product, which can compromise the quality and go against the new profile of consumers who are seeking for natural products aiming to have a healthier diet (Michulec & Wardenki, 2005).

The process known as pre-press is designed to prepare the high oil content raw material to the solvent extraction. Mechanical pressing reduces the oil by half to two thirds of its original level facilitating solvent extraction by reducing the amount of oil to be extracted, the size of extractor equipment and also the volume of solvent used (Walkelyn & Wan, 2006).

The principle of the method is to perform successive washing of oleaginous material with solvent until equilibrium is reached or near equilibrium between the oil content of the solid and that of the solvent, i.e., when the solvent absorbed by the solid and solvent in the free miscella containing the same amount of oil dissolved. When this occurs, the miscella is drained and another washing takes place.

The choice of solvent takes into account a number of factors, notably, solvent extraction capacity, effects of solvent on oil properties, process safety, solvent volatility and stability, and economic considerations (Attah & Ibemesi, 1990).

In 1947, it was published a study evaluating the various types of solvents such as benzene, aviation gasoline, methanol, ethanol, isopropanol, carbon disulfide, diethyl ether, ethylenedichloride, carbontetrachloride, trichloroethylene and the various petroleum naphthas. However, in 1947, the most commonly used solvents in the United States were light paraffinic petroleum fractions, such as the hexanes, heptanes and pentanes. The hexane was finally chosen because of its ease to evaporate and left no residual obnoxious odors or tastes. There are two types of hexane: normal, also known as n-hexane; and commercial, named extraction-grade hexane. The n-hexane is pure and boils at 69°C, while the extraction-grade hexane is not pure. The Clean Air Act of 1990 (public law) considered hexane as one of 189 hazardous air pollutants, encouraging research to seek alternative solvents for the extraction of oil. Examples of alternative solvents studied are halogenated solvents, water as a solvent, enzyme-assisted aqueous extraction, acetone as a solvent, alcohols as solvents and supercritical solvents (Williams, 2005).

The ability to extract an oil from an oilseed depends on the nature of the oil, the nature of the solvent, the temperature and the contact time between solvent and grain mass, flake thickness and conditions of pretreatment of the seeds. Attah & Ibemesi (1990) evaluated the solvent capacity and solvent effects in the extraction of oils from four native plants. The following solvents were used: petroleum benzene (60-80 °C), cyclohexane, isopropyl ether, ethyl acetate, tetrahydrofuran, propan-2-ol and acetone. The authors found that oil extraction performance of each solvent appears to be generally dependent on the nature of the oil, *i.e.* ethyl acetate and tetrahydrofuran gave the highest oil yields in rubber, their oil yields in melon gave next to lowest value. Another important consideration was that solvent did not affect the increase in the level of free fatty acid of extracted oils.

2.4 Supercritical CO₂ extraction

Although solvent extraction is efficient and has high oil yield, the problems presented by this technique have made scientists to research for new routes. One technique that has been used is the supercritical-fluid (SCF) extraction. A fluid is considered supercritical when it presents diffusivity similar to gas and density comparable to liquids.

Cagniard de la Tour had discovered the critical point of substances in the 1820s when it was observed the disappearance of the gas-liquid meniscus at temperatures higher than critical values under pressure, and in 1879 the solvent power of SCFs was reported. Since the 1930s, several potential applications of SCFs and liquefied gases had been proposed for extraction and separation processes, and purification of fatty oils. The technology only was brought into commercial focus in the 1960s by the work of Zosel at the Max-Planck Institute in Mannheim, Germany, which eventually led to its commercialization for coffee decaffeination and hops extraction purposes in Germany in the late 70s and early 80s (Temello & Guçlu-Ustundag, 2005).

In the 1980s, the use of supercritical CO_2 for the extraction of oilseeds such as soybean, cottonseed, corn germ, rapeseed, and sunflower. The oils obtained were shown to have similar quality compared with hexane extracts; they also had lighter color and lower iron and phospholipid content, resulting in a lower refining loss and reduction of subsequent refining steps. Another advantage is related to protein quality of the extracted meal that can be comparable with that of hexane-extracted (Temello & Guçlu-Ustundag, 2005).

Despite the many advantages of $SCCO_2$ extraction and also the high volume of research carried out, commercial-scale $SCCO_2$ extraction of oilseeds was not readily accepted. The process based on $SCCO_2$ extraction is simpler than conventional hexane extraction in terms of eliminating the need for hexane evaporators and meal desolventizer, but $SCCO_2$ process has some disadvantages such as high equipment costs and the inability to achieve continuous processing of high volumes of oilseeds under SCF conditions. Those two reasons are considered major impediments to commercialization of the SCCO$_2$ process. However, new research and developments in equipment design (i.e., the coupling of CO$_2$ with expeller technology) and government regulations on the use of hexane may render this system commercially possible in the near future (King, 1997 in Temello & Guçlu-Ustundag, 2005).

The main motivation for commercialization of SCF technology is the concern over the use of organic solvents, which was reflected in government regulations (in Germany the use of methylene chloride is forbidden) as well as changes in consumer attitude, who demands a product with high quality and also "natural". Other advantages are mild operating conditions for heatsensitive compounds (compared with distillation), and a solvent-free extract and residue (compared with solvent extraction). In addition, it provides an oxygenfree environment and limits oxidative degradation of the product (Temello & Guçlu-Ustundag, 2005).

Pradhan et al. (2010) compared supercritical CO$_2$ extraction method with conventional ones, such as solvent extraction and mechanical pressing, for flax seed. Supercritical-fluid extraction system used is shown in Figure 5.

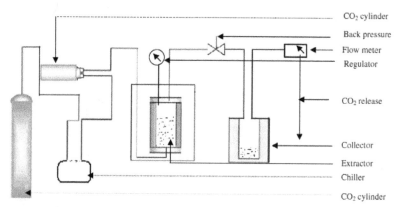

Fig. 5. Supercritical CO$_2$ apparatus (Pradhan et al., 2010).

The working principle of this system is presented briefly. A diaphragm compressor was used to compress the CO$_2$ to the desired pressure. Grains were placed in the preheated vessel and then the extraction with CO$_2$ began at a rate of 40 g / min for 3 hours. The extraction was held at 50 °C and pressure of 30MPa. Extracts were collected in another vial attached to the depressurization valve and cooled in a bath maintained at 0 °C (Pradhan et al., 2010).

The results showed that the oil yields were 38.8% to solvent extraction, 35.3% for the supercritical extraction and 25.5% for expelling. In terms of extracting the components omega-6-fatty acid and omega-3-fatty acid, supercritical extraction was more efficient than solvent and expelling extractions. The chemical composition of oil obtained from expelling is similar to that obtained by supercritical extraction, although the yield was lower. Extraction with hexane resulted in a higher oil yield but the quality of oil was lower in terms of its acid value and peroxide values, as well as presenting a lower concentration of omega fatty acids. They also observed that the solvent extraction also removed some waxy components and that oil produced contains traces of solvent.

2.5 Comparison of processes

The three processes discussed in this chapter, solvent extraction, supercritical CO_2 and mechanical extraction (screw press and batch hydraulic pressing) can be compared and evaluated under three main parameters: environmental, economic and oil yield.

Today, the process most commonly used commercially is solvent extraction. This process has the advantage of high oil yield (over 98%) and the total domination of the technological process. However, it has as a great disadvantage, the fact of using chemical solvents. After oil extraction, the remaining solvent should have a proper disposal to prevent environmental damage and avoid additional costs to the process, as well having higher investment cost (*i.e.* costly equipments). Another serious problem is the removal of the solvent from vegetable oil, ensuring adequate levels that are not harmful to human health. Besides oil, the meal is obtained, a co-product of solvent extraction, which depending on the feedstock source is rich in proteins, which could be used for animal feed, if high levels of solvent aren´t present. If necessary, one more step could be done, removing the solvent from the meal so it can be used. This process is not recommended for oilseeds with high oil content in its composition. The energy consumed to operate a plant for solvent extraction is high, and requires skilled labor to deal with this complex operation.

The method of extraction by supercritical CO_2 is still relatively new and is not widely used in commercial scale for the extraction of vegetable oils. It has the advantage of presenting a pure oil, without the presence of the solvent used. The oil yield is close to that obtained via solvent extraction and quality, close to that obtained via screw press methods. Studies are still being made in order to optimize the process and reduce costs. Here also the energy consumed to operate the equipment is high, and due to the complexity of the system, skilled labor is necessary.

Mechanical methods are the oldest methods of oil extraction. The great advantage of these methods is not using any kind of chemical products, producing a crude oil with high quality and ready for consumption, in some cases. Other important advantages are the low cost of acquisition of equipment, low power consumption for the operations and manpower do not need to be skilled. Despite the environmental advantages, the economics are unfavorable, since although the facilities have lower costs than those in the chemical process, the low oil yield, and high oil content in the cake, can make the process unprofitable.

3. Detailing the continuous mechanical pressing

3.1 Operating principle

The continuous mechanical pressing or simply screw press is shown in Figure 6. The process begins by putting the oilseeds inside the feed hopper. The screw press has a horizontal main shaft that carries the screw composition which is formed integrally with the shaft. The rotation of the screw occurs inside the cage or barrel that is a structure formed by steel bars. Spacers are placed between the lining bars allowing the drainage of oil as the pressure over the grains is increased. A movable cone or choke control is installed at the discharged end. This device has the function of operating pressure by changing the width of the annular space through which cake must pass. It is possible to adjust the choke by a hand-wheel on the opposite end of the screw (Khan & Hanna, 1983). Some machines have a device capable of removing heat generated by friction of the grains, making use of cold water circulation.

Both choke size and axis rotation speed should be adjusted when pressing different kinds of seeds (Jariene et al., 2008).

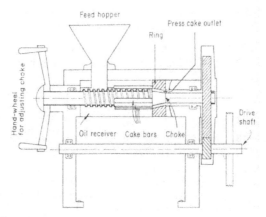

Fig. 6. Screw press (Khan & Hanna, 1983).

The working principle of the continuous mechanical pressing is to force the oilseed mass through the barrel by the action of the revolving worms. The volume of the mass is being reduced as the transition takes place through the barrel, causing compression of the cake and the resulting output of oil by the perforation of the lining bars of the barrel while the de-oiled cake is discharged through the annular orifice (Akinoso et al., 2009).

Screw presses can be powered with electric motors, diesel or even be operated manually (Jariene et al. 2008). Some researchers mentioned the application of photovoltaic cells to power screw presses (Mpagalile et al., 2007), as will be detailed below.

An important parameter related to the pressing efficiency is the determination of residual oil in the cake. High pressures can lead to cakes with less than 10% of oil content, which leads to higher crude oil production. A reduction in speed of rotation of the shaft, for example, can reduce the oil yield, increasing the oil content in the cake and solids in the oil (Jariene et al., 2008).

After pressing, the crude oil must be purified. A very common method of purification, especially for small producers, is to allow the oil to stand undisturbed for a few days, removing the upper layer (purified oil) and discard the bottom layer, consisting of fine pulp, water and resins. If this step is not enough, one can filter the oil using a press filter or using a centrifuge, but both methods are costly (Bachmann, 2004).

In addition to obtaining the oil, mechanical expelling produces a very important by-product named cake or meal. Some oilseeds cakes have high nutritional value and can be used as human food. Some are not suitable as food, but serve to complement the diet of chickens, pigs and cattle. It is important to emphasize the need for proper storage of seeds and cakes. They must be protected against the action of moisture, rodents and insects. High moisture content will generate mold, which can alter the taste of cakes, being rejected by the animals. Another problem is the development of mycotoxins such as aflatoxin, which in high concentrations can be poisonous to humans and animals. (Bachmann, 2004).

3.2 Optimization studies

According to Bargale (1997), despite the many advantages of mechanical extraction as previously mentioned, the low oil yield is still a limiting factor. In his thesis Bargale mentioned that US$ 57 million of edible oil are annually left in the deoiled cake because pressing extraction efficiencies rarely exceed 80%, compared to more than 98% obtained by solvent extraction. So, many studies are being conducted to increase the oil yield and also to perform the optimization of the process, defining the best values for the process variables, like applied pressure and axis rotation speed, as well as raw materials preparation by a number of unit operation like cleaning, heating, decorticating, cracking, flaking, cooking, extruding and drying. Khan & Hanna (1983) add to this list heating time, heating temperature, moisture content and pressing time, as variables that affect the oil yield during expression.

Ward (1976) reported that cooking and drying are the factors which most affect the performance of the screw press. Heating before pressing increases oil yield due to the breakdown of oil cells, coagulation of protein, adjustment of moisture content to the optimal value for pressing, and decreased oil viscosity, which allows the oil to flow more quickly.

Bamgboye & Adejumo (2007) developed a screw press to extract oil from decorticated (dehulled) sunflower seeds. The main components of their equipment are frame, cake outlet, expeller housing, heating compartment, auger, hopper, auger pulley and shaft. The seeds are steam heated by a heating compartment and the cake outlet is located at the end of the equipment where the seeds are compressed and the oil is forced out of the grains. Equipment performance was evaluated by testing three speeds of the axis (30, 40 and 50 rpm) and three levels of throughputs (1, 2 and 3), representing the number of times the material passes through the machine. The best oil yield (73.08%) was obtained for 50 rpm speed and 3 throughput of the cake. The lower oil yield was obtained for the conditions of 30 rpm and 3 throughput of the cake. The authors observed that oil yield increases with the increase in screw speed and throughput.

Pradhan et al. (2011) studied the effects of cooking (i.e., heating by oven) and moisture content on pressing characteristics of dehulled Jatropha seeds. The effect of moisture content (7.22, 9.69 and 12.16% wb – wet basis), cooking temperature (50, 70, 90, 110 and 130 ºC) and cooking time (5, 10, 15 and 20 min.) on parameters such as oil recovery, residual oil content, pressing rate and sediment content were evaluated. Oil recovery was affected by moisture content: as its increase leads to an increase in oil recovery. For cooked seeds, the highest oil yield was obtained for 8.19 % (db – dry basis) while for uncooked seeds, 9.86 % (db). Subsequent increases in moisture content resulted in a rapid drop in oil yield, probably due to mucilage development on oil cells that makes difficult the oil flow. The best cooking temperature was 110 ºC during 10 minutes. The first conclusion was that oil recovery from uncooked samples was lower than the cooked samples at same moisture content. When the seeds are cooked their tissues become softer and the oil has a lower viscosity; softer tissues tend to weaken the cellular structure, causing a rupture under pressure, while the low viscosity facilitates the flow of oil, contributing to higher oil yields. Higher temperatures can lead to a brittle seeds and oil degradation. In seeds with a high-protein content such as Jatropha seeds, protein coagulation due to heat treatment may have had a significant effect on oil recovery. The pressing rate from uncooked seed was higher than that from cooked seed and the sediment content of screw press oil was lower.

Effects of processing parameters on oil yield of finely and coarsely ground Roselle seeds were studied (Bamgboye & Adejumo, 2011). The study variables were the pressure applied during pressing, heating temperature of the grains, pressing time and heating time. Both processing parameters and size of the material affected the oil yield, which increased with an increase in the processing parameters of pressure up to 30 MPa and temperature of 100 ºC and decreased beyond these points. Also it can be observed that oil yield increases with an increase in moisture content. Finely ground samples showed a higher oil yield for different process parameters.

Deli et al. (2011) reported the influence of three types of accessory factors (nozzle size - the size of the choke section, speed and shaft screw press diameter) of the screw machines on the extraction of *Nigella sativa* L. seeds oils. The highest oil yields, 22.27% and 19.2%, were obtained for nozzle size of 6 mm, shaft screw diameter of 8 mm and a rotational speed of 21 rpm and 54 rpm, respectively. The lowest oil yield, 8.73%, was obtained for nozzle size of 10 mm, diameter of shaft screw of 11 mm and rotational speed of 21 rpm. The percentage of oil yield was decreased with the increase on the speed of the machine. When a slow rotation was used an increase in the pressing process time and in the heating of the grain mass was observed. That heating allows the oil to flow more easily and so, more oil is expelled. Nozzle size is directly related to applied pressure. Small openings add pressure to the seeds, thereby providing a higher temperature of the mass of grains due to friction between shaft screws and the seeds. As conclusions, the authors state that the oil yield increases with decreasing nozzle sizes, diameter of screw, speed and temperature.

The effects of compressive stress, feeding rate and speed of shaft screw press on palm kernel oil yield were evaluated by Akinoso et al. (2009). They used an expeller with a rated capacity of 180 kg/h. The experiments were conducted by using a factorial experimental design with 3 variables at 3 levels: compressive stress (10, 20 and 30 MPa), feeding rate (50, 100 and 150 kg/h) and shaft screw press speed (50, 80 and 110 rpm). A maximum efficiency, 94.5 %, was obtained at 30 MPa compressive stress, 150 kg/h feed rate and 110 rpm of speed of shaft screw press. A minimum express efficiency, 33.6 %, was obtained at 10 MPa compressive stress, 150 kg/h feed rate and 50 rpm of shaft screw press speed. For the studied range, oil yield increases with increase in speed of shaft screw press and in feeding rate. Oil yield is directly proportional to compressive stress while influence of speed is marginal, and also it is possible to predict a further increase in oil yield with an increase in compressive stress.

Evangelista & Cermak (2007) studied the effects of moisture content of cooked flaked seeds of *Cuphea* (PSR23) on continuous screw pressing characteristics of the seed and quality of crude press oil. Press cake analysis showed that there was a significant decrease in the residual oil content when the moisture content was reduced. For moisture content of 9 and 12%, the difference in residual oil content was not significant. There was a decrease of 0.4% in the residual oil present in the cake when the moisture content was reduced by 1%. The lowest value of residual oil in the cake, 5.6%, was achieved at 3.1% of moisture content. Oil yield ranged from 79.4 to 83.6% as the cooked seed moisture content decreased from 5.5 to 3.1%. Despite the dry seeds at 3% present the best oil yield, the authors highlighted that long periods of drying and high pressures applied increase energy consumption of the process.

Olayanju (2003) studied the effects of speed of shaft screw press (30 to 75 rpm) and moisture content (4.10 to 10.32 % wb) on oil cake qualities of expelled sesame seed. As conclusions, the author found that the color of the oil was darkened as the speed increased from 30 to 75 rpm and with higher initial moisture content. With a higher moisture content, the residual oil in the cake increased. The lowest residual oil in cake, and hence, the highest oil yield, was obtained at 45 rpm and moisture content of 5.3%.

Screw pressing of crambe seeds was studied by Singh et al. (2002). They evaluated the influence of moisture content and cooking on oil recovery. For cooked seeds, the oil yield increased from 69 to 80.9% with the decrease in moisture content, while for the uncooked seeds, the oil yield increase was from 67.7 to 78.9%. Low moisture content, i.e. 3.6%, resulted in plugging of the screw press. Cooked seeds had higher oil recovery (7%) than uncooked (3.6%). They also concluded that drying was much more beneficial than cooking in terms of oil recovery for the range of conditions in their study but cooking is highly recommended in case of crambe to inactivate the enzyme myrosinase, making the cake suitable for livestock feed.

4. Examples on the application of pressing for obtaining oil from cotton, peanut and sunflower

In this last part of the chapter some practical examples are shown. Results were obtained from the author's thesis developed at the Campinas State University. The objective of this work was to evaluate the production of vegetable oils, i.e. cotton, peanuts and sunflower, for biodiesel production. The process adopted is small screw press with capacity of 40 kg of oilseeds per hour. For a better evaluation of the process and results, experimental design and the methodology of surface response was chosen to use to analyze the results. The parameters chosen to evaluate the oil yield were shaft screw press speed, moisture content and temperature of the grains. Based on the lipid content found for each grain, it was possible to calculate the mass of oil corresponding to 5 kg of kernels that was pressed and so, the oil yield.

4.1 Peanut (*Arachis hypogaea* L.)

The peanuts used in the experiments have 18.25% of hulls and 81.75% of grains, with a moisture content of 6.20% and lipid content, determined only for the grain of 39%. The experiments were conducted with the grain in its initial moisture content (6.20% wb) and with hulls. Table 2 shows the yields of crude oil, obtained experimentally.

For higher temperatures and smaller speeds, yields are higher. In the speed range between 80 and 90 rpm, the oil yield tended to increase, regardless of the temperature range used. What was observed in the experimental design was confirmed by observations during pressing. At lower screw press speeds, the grains had more time inside the press and the contact of grain with an additional heating provided by the equipment favors oil expelling. High speeds render difficult to the press to crush the grains properly, undermining the elimination of the oil from inside the seeds. It is important to note that there is a minimum permissible speed, which was the one used in this work. Values less than this minimum generate an increase in particulates in the oil.

No.	Speed (rpm)	Temperature (ºC)	Oil yield (%)
1	90	72	54.67
2	90	107	45.49
3	114	72	46.67
4	114	107	25.90
5	85	85	61.69
6	119	85	44.41
7	102	60	23.74
8	102	110	40.26
9	102	85	32.36
10	102	85	34.56

Table 2. Oil yields from peanut (*Arachis hypogaea* L.)

Regarding the screw press speeds, a value between 80 and 90 rpm showed the better performance in oil yield and also at this range it was possible to observe an improved performance of the press, producing a cake with a proper consistency and oil with no particulates. Tests 7 to 10 showed that for the same speed an increase in temperature leads to increase in oil yield. However, the temperature factor was not considered significant from a statistical viewpoint. Although temperature, statistically did not influence the oil yield, it was observed that the press worked better with the grains heated, reducing the pressing time and the press operation cost. As the heating step is costly, it is recommended to heat the grains at lower temperatures, between 40 and 50 ºC. The moisture content which favored the press, without contaminating the oil with water, is within the range of peanut commercialization (8 and 12 % wb), thus not requiring any step prior to pressing, such as drying.

4.2 Cottonseed (*Gossypium* L.)

The cottonseed with lint has moisture content of 5.58 % (wb) and 16.9 % of lipid content. Data analysis according to the methodology of surface response indicated that none of the studied variables significantly affected the oil yield of cottonseed. However, based on the experiments it was possible to define an operating range suitable for processing this type of cottonseed in small equipment. The proposed operating conditions for better pressing of cottonseed was 85 rpm speed of shaft screw press, 9 % (wb) moisture content, the same of commercialized cottonseed, and temperatures between 110 and 120 ºC.

4.3 Sunflower (*Helianthus annuus* L.)

The sunflower seeds used in the pressing had an initial moisture content of 7.5% (wb) and 46.8% of lipid content. During preliminary tests of sunflower pressing it was observed that grains with moisture contents other than 7.5% did not have good performance. For this reason, the experimental design took into account only the temperature and the speed of shaft screw press, which proved to be the most important factors. The results are shown in Table 3.

The highest yield, 67%, was achieved for 90 rpm speed and 72 ºC grain temperature, while the lowest yield, 48.81%, was obtained at the speed of 114 rpm and of 72 ºC. During the

No.	Speed (rpm)	Temperature (°C)	Oil yield (%)
1	90	72	67.28
2	90	107	53.96
3	114	72	48.81
4	114	107	59.45
5	85	85	67.23
6	119	85	62.17
7	102	60	66.89
8	102	110	54.94
9	102	85	66.60
10	102	85	65.45

Table 3. Oil yields from sunflower (*Helianthus annuus* L.)

experiments it was observed that at higher RPMs, the oil yield attained was higher even when the temperature was kept constant. Since the statistical analysis showed that for lower values of temperature the oil yield was higher, a new experimental design was proposed with another temperature range (Table 4).

No.	Speed (rpm)	Temperature (°C)	Oil yield (%)
1	90	30	54.77
2	90	55	61.49
3	114	30	68.38
4	114	55	64.38
5	85	42.5	51.70
6	119	42.5	62.72
7	102	25	61.11
8	102	60	64.85
9	102	42.5	68.04
10	102	42.5	66.13

Table 4. Oil yields from sunflower (*Helianthus annuus* L.) – new design.

The analysis of all experimental data obtained for sunflower oil expelling help to define value ranges for each operating variables. Regarding the screw press speeds, sunflower requires higher RPMs than the peanuts and cottonseed, around 100 to 115 rpm. This difference can be explained by the geometry of the grain and the absence of a significant element of friction, such as the hulls of peanuts and cottonseed lint. During the pressings it was possible to observe how the operation was better conducted at faster speeds, allowing a shorter pressing time. Temperatures close to room temperature (25 and 30 °C) are appropriate, leading to energy saving in the process.

5. Conclusion

Vegetable oils are of great importance to human health as well as to the development of oil chemistry. It can be produced from many technological processes, chemical and mechanical.

Solvent extraction technology is still the most widely used process and the one with higher oil yield, but due to the use of chemicals which affect the environment and human health, new technologies have been researched, such as supercritical extraction. Methods not as new as supercritical extraction, such as screw press, have been the object of study, in order to optimize this simple process so it can produce more oil, with high quality at a lower cost. Several studies focusing on a wide range of oilseeds and technologies, all of them saying the same thing. An extraction process should provide economic and environmental advantages to become an winner technology.

6. Acknowledgment

We acknowledge the EMBRAPA AGROENERGY, Faculty of Agricultural Engineering (FEAGRI/UNICAMP)), the Food Technology Institute (ITAL), the State of São Paulo Research Foundation (FAPESP) for its financial and technical support. Also, we would like to acknowledge DSc. José Manuel Cabral de Sousa Dias and MSc. Larissa Andreani for their suggestions and revision in the text.

7. References

Anderson, D (2005). A primer on oils processing technology, In: *Bailey's Industrial Oil and Fat products*, F. Shahidi (Ed), Vol. 5, 1-9, ISBN 0-471-38552-2, New Jersey: John Wiley & Sons, Inc.

Akinoso, R.; Raji, A. O.; Igbeka, J. C. Effects of compressive stress, feeding rate and speed of rotation on palm kernel oil yield. *Journal of Food Engineering*, Vol. 93 (2009), pp. 427-430, ISSN 0260-8774.

Attah, J. C.; Ibemesi, J. A. Solvent extraction of the oils of rubber, melon, pumpkin and oilbean seeds. *Journal of the American Oil Chemist's Society*, Vol. 67 (1990), pp. 25-27, ISSN 1558-9331.

Bachmann, J. Oil seed processing for small scale producers. National sustainable agriculture information service (2004). Available in: <http://www.attra.ncat.org/attra-pub/PDF/oilseed.pdf>. Acessed 01 july 2011.

Bamgboye, A. I.; Adejumo, O. I. Effects of processing parameters of Roselle seed on its oil yield. *International Journal of Agricultural and Biological Engineering*, Vol. 4 (2011), pp. 82-86, ISSN 1934-6352.

Bamgboye, A. I.; Adejumo, A. O. D. (2007) Development of a sunflower oil expeller. *Agricultural Engineering International: the CIGR Ejournal*. Manuscript EE 06 015, vol IX. September.

Bargale, P. C. Mechanical oil expression from selected oilseeds under uniaxial compression. PhD Thesis, Department of Agriculture and Bioresource Engineering, University of Saskatchewan, Saskatoon, 1997.

Behr, A.; Gomes, J. P. The refinement of renewable resources: New important derivatives of fatty acids and glycerol. *European Journal of Lipid Science Technology*, Vol. 112 (2010), pp. 31-50, ISSN 1438-9312.

Deli, S.; Masturah, M. F.; Aris, Y. T.; Nadiah, W. A. W. The effects of physical parameters of the screw press oil expeller on oil yield from *Nigella sativa* L seeds. *International Food Research Journal*, Vol. 18 (2011), pp. 1367-1373, ISSN 2231-7546.

Evangelista, R. L.; Cermak, S. C. Full-press oil extraction of *Cuphea* (PSR23) seeds. *Journal of the American Oil Chemistry's Society*, Vol. 84 (2007), pp. 1169-1175, ISSN 1558-9331.

Gunstone, F. D. (2005). Vegetable oils, In: *Bailey's Industrial Oil and Fat products*, F. Shahidi (Ed), Vol. 1, 213-267, ISBN 0-471-38552-2, New Jersey: John Wiley & Sons, Inc.

Jariene, E.; Danilcenko, H.; Aleknviciene, P.; Kulaitiene, J (2008). Expression – Extraction of pumpkin oil, In: *Experiments in unit operations and processing of foods*, M. M. C. Vieira; P. Ho (Eds), 53-61, ISBN 978-0-387-68642-4, Iceland: Springer.

Kemper, T. G (2005). Oil Extraction, In: *Bailey's Industrial Oil and Fat products*, F. Shahidi (Ed), Vol. 5, 63-68, ISBN 0-471-38552-2, New Jersey: John Wiley & Sons, Inc.

Khan, L. M.; Hanna, M. A. Expression of oil from oilseeds – A Review. *Journal of Agricultural Engineering Research*, Vol. 28 (1983), pp. 495-503, ISBN 1095-9246.

King, J. W (1997). Technology and solvents for extracting oilseeds and nonpetroleum oils, In: P. J.Wan and P. J.Wakelyn (Eds), p. 283, Champaign: AOCS Press.

Koo, E. C. Expression of vegetable oils. *Industrial and Engineering Chemistry*, Vol. 34 (1942), pp. 342–345, ISSN 0888-5885.

Mpagalile, J. J.; Hanna, M. A.; Weber, R. Seed oil extraction using a solar powered screw press. *Industrial Crops and Products*, Vol. 25 (2007), pp. 101-107, ISSN 0926-6690.

Michulec, M.; Wardencki, W. Development of headspace solid-phase microextraction–gas chromatography method for the determination of solvent residues in edible oils and pharmaceuticals. *Journal of Chromatography A*, Vol. 1071 (2005), pp. 119–124, ISSN 0021-9673.

Mrema, G. C.; McNulty, P. B. Mathematical-model of mechanical oil expression from oilseeds. *Journal of Agricultural Engineering Research*, Vol. 31 (1985), pp. 361–370, ISSN 1095-9246.

Olayanju, T. M. A. Effect of wormshaft speed and moisture content on oil and cake qualities of expelled sesame seed. *Tropical Science*, Vol. 43 (2003), pp. 181-183, ISSN 1556-9179.

Owolarafe, O. K.; Osunlekeb, A.S.; Odejobic, O.A.; Ajadid, S.O.; Faborodea, M.O. Mathematical modelling and simulation of the hydraulic expression of oil from oil palm fruit. *Biosystems Engineering*, Vol. 101 (2008), pp. 331-340, ISSN 1537-5129.

Pighinelli, A.L.M.T. Study of mechanical expeller and ethanolic transesterification of vegetable oils. PhD Thesis, School of Agricultural Engineering, Campinas State University (UNICAMP), Campinas, 2010.

Pradhan, R. C.; Mishra, S.; Naik, S. N.; Bhatnagar, N.; Vijay, V. K. Oil expression from Jatropha seeds using a screw press expeller. *Biosystems Engineering*, Vol. 109 (2011), pp. 158-166, ISSN 1537-5129.

Pradhan, R.C.; Meda, V.; Rout, P.K.; Naik, S.; Dalai, A. K. Supercritical CO_2 extraction of fatty oil from flaxseed and comparison with screw press expression and solvent extraction processes. *Journal of Food Engineering*, Vol. 98 (2010), pp. 393-397, ISSN 0260-8774.

Singh, K. K.; Wiesenborn, D. P.; Tostenson, K.; Kangas, N. Influence of moisture content and cooking on screw pressing of crambe seed. *Journal of the American Oil Chemist's Society*, Vol. 79 (2002), pp. 165-170, ISSN 1558-9331.

Temello, F.; Guçlu-Ustundag, O (2005). Supercritical technologies for further processing of edible oils, In: *Bailey's Industrial Oil and Fat products*, F. Shahidi (Ed), Vol. 5, 397-432, ISBN 0-471-38552-2, New Jersey: John Wiley & Sons, Inc.

Walkelyn, P.J.; Wan, P.J. (2006). Solvent extraction to obtain edible oil products, In: *Handbook of Functional Lipids*, C.C. Akoh (Ed), 89-131, ISBN 978-0-8493-2162-7, Boca Raton: CRC Press.

Ward, J. A. Processing high oil content seeds in continuous screw presses. *Journal of the American Oil Chemists´ Society*, Vol. 53 (1976), pp. 261-264, ISSN 1558-9331.

Weiss, E. A. (1983). Oilseed processing and products, In: *Oilseed crops*, 528-596, ISBN 0-632-05259-7, London: Longman.

Willems, P.; Kuipers, N. J. M; De Haan, A. B. Hydraulic pressing of oilseeds: Experimental determination and modeling of yield and pressing rates. *Journal of Food Engineering*, Vol. 89 (2008), pp. 8-16, ISSN 0260-8774.

Williams, M.A. (2005). Recovery of oils and fats from oilseeds and fatty materials, In: *Bailey´s Industrial Oil and Fat products*, F. Shahidi (Ed), Vol. 5, 99-189, ISBN 0-471-38552-2, New Jersey: John Wiley & Sons, Inc.

World Bank Group. Vegetable oil processing (1998). Available in:
    <http://www.ifc.org/ifcext/enviro.nsf/AttachmentsByTitle/gui_vegoil_WB/$FILE/vegoil_PPAH.pdf>. Acessed 01 august 2011.

4

Nitrogen Efficiency in Oilseed Rape and Its Physiological Mechanism

Zhen-hua Zhang[1,2,3,4], Hai-xing Song[1,2,3,4]* and Chunyun Guan[5]
*1Department of Plant Nutrition,
College of Resources and Environment,
Hunan Agricultural University,
2Hunan Provincial Key Laboratory of Plant Nutrition in
Common University, Changsha,
3Hunan Provincial Key Laboratories of Farmland Pollution Control and
Agricultural Resources Use, Changsha,
4National Engineering Laboratory of High Efficiency Utilization of Soil and
Fertilizer Resoureces, Changsha,
5National Center of Oilseed Crops Improvement,
Hunan Branch, Changsha,
China*

1. Introduction

N fertilizer application can guarantee the high yield of crop; it is a general method to improve the yield of crop (Zhang et al., 2010). However, not only N use efficiency was declined, but also environment contamination was serious day by day because of N fertilizer over application (Zhang et al., 2010). So, control and reduction in the amount of N fertilizer application is very important, it is necessary to dredge up the potential of N absroption and N use efficiency of crop. Oilseed rape (Brassica napus L.) is an important oil production crop in China. However, the N use efficiency and N efficiency of oilseed rape are very low (Schjoerring et al., 1995). The N application amount has reached 200-330kgN per ha (Schjoerring et al., 1995; Wiesler et al., 2001a) and it is increasing year by year. But the studies on differences of N efficiency in different oilseed rape varieties and breeding research of the oilseed rape with high N efficiency were relatively slower than other cereal (Liu et al., 2009). Oilseed rape requires high amounts of N for growth, but the N efficiency (seed yield per unit of accumulated N in plant) is very low. Consequently, it is necessary to improve the N efficiency (Rathke et al., 2006). In generally, N efficiency has two components: N uptake efficiency and N utilization efficiency (Sattelmacher et al., 1994). The differences of N efficiency between oilseed rape genotypes are significant (Wiesler et al., 2001b; Christian et al., 1999). Kessel et al., 1999 has been measured N concentration of organs in different oilseed rape genotypes, results suggested that oilseed rape possess the 2 physiological characteristics (low N concentration in dropped leaves, high N harvesting

---

* Corresponding Author

index) has higher N efficiency also. The development of N efficient genotypes and improvement of N management will require to understand the relationship between physiological processes and biomass, yield formation of crop under no N application conditions (Dreccer et al., 2000). Mahmoud et al., (2005) compared the differences of root growth and nitrate N exhaust in cultivar culture between high N efficiency and low N efficiency oilseed rape; larger amount of root biomass, higher root uptake activity and higher exhausted nitrate N amount were found in the high N efficiency oilseed rape; Nitrate-N uptake from soil depends on root growth and uptake activity, the amount of N depleted from the compartments significantly correlated with root-length density (Mahmoud et al., 2005). Seiffert et al., (2004) suggested that, the N use efficiency can be increased significantly by strengthen the activities of asparagine synthetase and glutamine synthetase through transgenic methods. Obviously, it is necessary to 'system study' on N uptake and N use efficiency in different oilseed rape. The differences of yield, N absorption and N use efficiency in different oilseed rape genotypes should be compared, and preliminarily discussion the contribution of N absorption efficiency and N use efficiency to N efficiency under different N application levels, in order to supply scientific basic and plant materials for the future study.

The differences of concentration and distribution of N in crop is depending on the differences of organs and growth stages, and N redistribution in different organs will occur at different growth stages; these differences are related to the transfer of growth center (Peoples et al., 1998; Zhang et al., 2008). Studies in wheat (Andersson et al., 2005), oilseed rape (Malagoli et al., 2005) and pea (Séverine et al., 2005) showed that the leaves and stems of crop have become important N "sources" after the flowering stage, grain N does not only come from root N uptake during the later growing period, but also from that redistributed from the vegetative organs (Zhang et al., 2010). The latter was one of the key factors that guarantee crop N requirement during the later growing period. Nutrient competition between different organs of the crop during the later growing period is fierce; the uptake ability of the root declines at this time and soil mineral nutrients are exhausted (Dong et al., 2009). In oilseed rape, all leaves drop off at harvesting stage and N in the leaves is lost. To achieve high yield and high N use efficiency in crop production, N must be redistributed from the roots to the grains, residue N in uneconomic tissues must be reduced, and N harvesting index of economic tissues must be increased (Zhang et al., 2010). Research on N redistribution in cereals reached the following conclusions. The redistribution proportion of N from the leaves, stems and reproductive organs to the grains was more than 60% in wheat (Palta et al., 1995a, 1995b). Dalling et al., (1985) showed that, although the soil can supply adequate N after the crop flowering stage, at least 50% of the N in grains is still redistributed from the vegetative organs. The effect of environmental conditions on N redistribution was considerable e.g., low N in barley (Przulj et al., 2001) and drought stress in wheat (Barbottin et al., 2005; Palta et al., 1994) can accelerate N redistribution in plants, but the process can be restrained when plants are damaged by diseases and insects e.g., wheat (Dimmock et al., 2002). However, only a few studies looked at N redistribution in oilseed rape. Malagoli et al., (2005) found that in oilseed rape the N of flowers and pods was mainly from endogenous N, comprising 64 and 73% of total N content in these tissues, respectively. The N requirements of seed filling were mainly satisfied by N mobilized from the vegetative parts. Rossato et al.,

(2001) studies of *Brassica napus* L. suggested that large amounts of endogenous N were redistributed from the leaves to the stems to the roots tap tissues which acted as a buffering storage compartment and later used to supply the reproductive tissues. About 15% of total N cycling through the plant was lost through leaf fall and 48% had been remobilized from vegetative tissues and finally recovered in the mature pods (Thomas et al., 2007). It can be concluded that, the accumulation and distribution characters of N in vegetative and reproductive organs are important factors that affected the crop production (Zhang et al., 2006), the higher yield production is not only dependent on higher N absorption, but also depend on higher N distribution efficiency (Wang et al., 2003). Studies on the regular of N absorption, distribution and redistribution at different growth stages will supply scientific foundation for rational application of N fertilizer, improving N harvesting index and increasing crop production. However, most of these studies mainly focused on the redistribution of N in cereal crops such as wheat (Anderson et al., 2005; Dalling et al., 1985), maize (Pommel et al., 2006) and rice (He et al., 2002). A few of them (Rossato et al., 2002; Malagoli et al., 2005) looked at oilseed rape. Moreover, studies on the relationship between N redistribution and crop production were fewer. In this study, two oilseed rape varieties (X-36 and X-50) with different N use efficiencies were grown in sand culture with complete nutrient solution and normal N supply. To better understand the N dynamics during later growth stages, the relationship between N redistributed from the vegetative organs to the grain and grain yield was studied using the ^{15}N labeling method.

2. Materials and methods

2.1 Comparison of Nitrogen (N) efficiency between oilseed rape genotypes

2.1.1 Plant materials and experiment design

The experiment was conducted at Agricultural Resources and Environmental College Experiment Field in HuNan Agricultural University during Sep. 2004 to May. 2005. 16 oilseed rape varieties were used as plant materials, the numbers are shown in Table 2, plant materials were supplied by China National Oilseed Crop Improvement Center, Hunan Branch. The soil used was an alluvial for vegetable cultivation derived from river flow alluvial material, containing organic matter 27.79 g/kg, total N 1.90 g/kg, and total P 0.79 g/kg, total K 19.76 g/kg. The NaOH hydrolyzed N was 108.31 mg/kg of soil, Olsen-P 15.85 mg/kg of soil, available K 19.76 g/kg of soil with pH 5.67. Urea was used as N fertilizer, calcium magnesium phosphate as P fertilizer (containing P_2O_5 12%) and potassium chloride (containing K_2O 60%) as K fertilizer.

The experiment had two N treatments: N application and no N application, 16 oilseed rape varieties used as plant materials, 32 treatments, 3 replicates, 96 districts, 10.5 m² per district, randomized block. N application treatment: 225kg N, 75kg P_2O_5 and150 kg K_2O per hectare, 50% N fertilizer used as basal fertilizer, 20% N fertilizer used as added fertilizer during winter, 30% N fertilizer used at stem elongation stage, all P and K fertilizer were used as basal fertilizer; no N application treatment: the same with N application treatment except of non-N fertilizer applied. Seedling on Sep. 25 2004, transplanted on Nov. 2 2004, transplanted density was 100 thousand plants per ha, management as normal.

2.1.2 Sampling, determination and calculation

Sampled whole plant at harvesting stage, separated it according to organs after clear, dried until constant weight, measured dry weight and total N. The measurement methods followed by the general agricultural chemistry analysis methods; the plant total N measured by Kjedahl method (Hardy et al., 1968). Below formula was used to calculate N use efficiency based on total N, N use efficiency=grain yield / plant total N. Data has been analysis by T-test using SPSS statistic software (Norusis, 1990), the significant differences between the two varieties (p<0.05) are figured by stars (*) in Table 6.

2.2 Vegetative N redistribution during later growing period and its contribution to N efficiency

2.2.1 Variety tested and crop culture

High N use efficiency variety X-36 and low N use efficiency variety X-50 were used as plant materials and grown in a glasshouse in the field experiment station of Hunan Agricultural University (Southern China) with average temperature 19 °C and natural light. These materials were selected from 30 varieties that have been evaluated in the last 10 years in Hunan Province. All the varieties were provided by the Hunan Sub-center of the Improvement Center of the National Oil Crop.

The sand culture and ^{15}N labeling method were used in this experiment. On a 30 cm × 30 cm brown plastic bowl, sand culture (water and dilute hydrochloric acid were used to clean the growth matrix) with complete nutrient solution and normal N supply was used as growth medium. The culture solution had these components: KNO_3 5 mmol $\cdot L^{-1}$, KH_2PO_4 1 mmol $\cdot L^{-1}$, $MgSO_4$ 7 mmol $\cdot L^{-1}$, $Ca(NO_3)_2 \cdot 4H_2O$ 5 mmol $\cdot L^{-1}$, Fe-EDTA 3 mmol $\cdot L^{-1}$, B 0.5 mg $\cdot L^{-1}$, Mn 0.5 mg $\cdot L^{-1}$, Zn 0.05 mg $\cdot L^{-1}$, Cu 0.02 mg $\cdot L^{-1}$, Mo 0.01 mg $\cdot L^{-1}$, $Ca(^{15}NO_3)_2$ and $K^{15}NO_3$ (Shanghai Chemical Engineering Corporation Research Institute) with ^{15}N excess = 20.28% were used as N sources for the labeling treatment.

2.2.2 Experimental treatment and labeling

The seedlings were planted on Sept 25, 2005; transplanted on Nov 2, 2005; one plant per bowl, 72 bowls per variety and harvested on May 5, 2006. ^{15}N labeled days and time of sampling are shown in Table 1. Labeling was done at the seedling stage (Dec 2-22, 2005; 21 d), stem elongation stage (Jan 21-25, 2006; 5 d), flowering stage (Mar 3-7, 2006; 5 d) and siliquing stage (Apr 15-23, 2006; 9 d) . Samples were taken at the end of each labeling stage (Dec 23, 2005; Jan 26, 2006; Mar 8, 2006; Apr 24, 2006). The other plants were transplanted into sand culture with no ^{15}N nutrient and sampled during later growth stages (flowering stage, Mar 8, 2006; siliquing stage, Apr 24, 2006; harvesting stage, May 5, 2006; Table 1), in order to distinguish between N redistributed and N taken up.

Labeling end stage in Table 1 means the end of each labeling treatment at four growth stages. Take seedling stage for instance. Labeling treatment was done at the seedling stage (Dec 2-22, 2005; 21 d); labeling treatment application on Dec 2, 2005; and finished on Dec 22, 2005. Sampling time at labeling end stage means 6 plants per variety were sampled on Dec 23, 2005. The other plants (12) were transplanted into sand culture with no ^{15}N nutrient and sampled at siliquing stage (Apr 24, 2006) and harvesting stage (May 5, 2006), in order to distinguish between N redistributed and N taken up. Sampling schedule for the other stages are shown in the text and this table.

Days for ^{15}N labeling	Pot number	Sampling time			
		Labeling end stage	Flowering stage	Siliquing stage	Harvesting stage
Seedling stage, 21 days	3×6=18	Root, stem, leaf		Root, stem, leaf, Silique	Root, stem, leaf, silique husk, grain
Stem elongation stage, 5 days	4×6=24	Root, stem, leaf	Root, Stem, Leaf, Flower	Root, stem, leaf, Silique	Root, stem, leaf, silique husk, grain
Flowering stage, 5 days	3×6=18	Root, stem, leaf, flower		Root, stem, leaf, silique	Root, stem, leaf, silique husk, grain
Siliquing stage, 9 days	2×6=12	Root, stem, leaf, silique			Root, stem, leaf, silique husk, grain

Table 1. Time of labeling (^{15}N) and sampling scheme at each growth stages

2.2.3 Sampling, determination and calculation

Whole plants were sampled according to the design indicated in Table 1. There were 6 replications (randomly selected were 6 plants per sampling time per variety), subsamples were taken from different tissues after clearing and drying. Biomass dried to a constant weight was measured. Dry samples were ground and sifted for isotope analysis. To measure biomass and calculate total N exactly, fallen leaves were collected in time. The concentration of N in the plant was measured using the Kjeldahl method (Kamprath et al., 1982); plant total N, grain total N, harvesting index and N harvesting index were calculated using the following formulas:

$$\text{Total N in plant} = \text{plant biomass} \times \text{N concentration in plant tissues}$$

$$\text{Grain total N} = \text{grain yield} \times \text{N concentration in grain}$$

$$\text{Harvesting index} = \text{grain yield/biomass}$$

$$\text{N harvesting index} = \text{grain total N/ total N in plant}$$

The samples were pre-processed by semi micro determination (Brenna et al., 1998). The excess ^{15}N in each tissue was measured by the use of a continuous flow isotope mass spectrograph at the Institute of Genetics, Hebei Academy of Agricultural Science. The amount and proportion of redistributed N from the vegetative organs to the reproductive organs were calculated as follows:

N redistribution proportion = ^{15}N accumulated amount in reproductive organs/^{15}N accumulated amount in the plant at the end of the labeling treatment×100 % (Because the flower and silique are reproductive organs of the plant, labeling was done at the flowering and siliquing stages, the ^{15}N accumulated amount in the reproductive organs should be subtracted from the ^{15}N accumulated amount in the reproductive organs at the end of the labeling treatment.);

N redistribution amount = N redistribution proportion × N accumulated amount in the plant at the end of labeling treatment;

N loss proportion = [^{15}N accumulated amount in the plant at the end of the labeling treatment - ^{15}N accumulated amount in the plant at harvesting stage (or siliquing stage)]/^{15}N accumulated amount in the plant at the end of labeling treatment×100 %;

N loss amount = proportion of N loss × N accumulated amount in the plant at the end of the labeling treatment

The partitioning of absorbed N was calculated from the excess ^{15}N in each tissue combined with the calculated total plant N uptake. This is based on the assumption that unlabeled N from the sand culture was taken up and allocated in the different plant tissues in a way similar to that of labeled N. All data were analyzed using the T-test by SPSS statistic software. The significant differences between the two varieties ($p<0.05$) were indicated by asterisks (*) in the table.

3. Results

3.1 Grain yield between different oilseed rape varieties

The results of grain yield showed that, the differences of grain yield between different oilseed rape varieties were significant, regardless N application level (Table2). The rang of yield variation in 16 varieties were 1060-1913 kg per hm^2 under no N application condition; the yields of Xy1, Xy7 and Xh20 were higher (≥2720kg per hm^2), the yields of Xy8, Xy9, Xy11

Variety No.	No N application	N application	N application - No N application
Xy1	1913a	2490ab	577bc
Xy6	1132de	1630e	498bcd
Xy7	1908a	2290bc	382cd
Xy8	1105e	1620e	515bcd
Xy9	1060e	1550e	490bcd
Xy11	1063e	1650e	587bc
Xy12	1730ab	2070cd	340d
Xy13	1381cd	2429abc	1048a
Xy14	1659ab	1770de	111e
Xy15	1742ab	1720de	-22e
Xy16	1696ab	2290bc	594bc
Xy17	1813ab	2445ab	632b
Xh19	1801ab	2720a	918a
Xh20	1908a	2786a	878a
Xy21	1837ab	2480ab	643b
Xy24	1588bc	1920de	332d

Different letters in the same row indicated that differences between varieties in t-test are significant at $p<0.05$ level. The same as below.

Table 2. Grain yield of different oilseed rape varieties

were lower (≤1105 kg per hm²). N response can be expressed by ratio of (yield with N application - yield with no N application) / N application amount. Because the N application amount of all varieties under N application treatment were the same in this study; so, the N response can be indirectly expressed by the differences of yield under N application condition and no N application condition. N responses of Xy13, Xh19 and Xh20 were higher than other varieties significantly; N responses of Xy14, Xy15 were lower than other varieties significantly. In the case of Xy15, grain yield can be increased by N application significantly, average increased of yield in sixteen cultivars were 35%.

3.2 N absorption amount and N use efficiency in different oilseed rape varieties

Results of table3 showed the relationship between N absorption amount and grain yield. N absorbed amount of different oilseed rape varieties were different under different N application level conditions. The differences of N absorption amount between varieties were significant under no N application condition; however, the differences were smaller under N application condition. N absorption amount of Xy1, Xy21 and Xy24 (≥1.688g per plant) were more than other varieties Xy8, Xy9 and Xy13 (≤1.293g per plant) significantly. The relationship between N absorption amount and grain yield showed that, the correlation between N absorption amount and grain yield was significant under no N application condition, correlation coefficient was 0.685** ($R_{0.01}$=0.623); but under N application condition, there was no significant correlation, and the correlation coefficient was 0.415($R_{0.05}$=0.497). This indicates that, only under the no N application condition, grain yield can be increased significantly by increased N absorption amount. The differences of N use efficiency between varieties were significant according to the differences of N absorption amount, regardless of N application level. The changes scope of N use efficiency in sixteen varieties was 7.1-12.6, the highest was Xh20 of 12.6, and the lowest was Xy9 of 7.9.

| Variety | N absorption (g/plant) | | N use efficiency (g/g) | |
No.	No N application	N application	No N application	N application
Xy1	1.744a	2.200a	11.0bc	11.4abc
Xy6	1.467d	2.207a	7.7ef	7.4e
Xy7	1.448d	2.209a	13.2a	10.4bc
Xy8	1.254e	2.151a	8.8def	7.6e
Xy9	1.285e	2.186a	8.2ef	7.1e
Xy11	1.456d	2.154a	7.3f	7.7e
Xy12	1.545bc	2.149a	11.2abc	9.6cd
Xy13	1.293e	2.245a	10.7bcd	10.8abc
Xy14	1.462d	2.192a	11.4abc	8.1de
Xy15	1.518cd	2.177a	11.5ab	7.9de
Xy16	1.489cd	2.177a	11.4abc	10.5bc
Xy17	1.500cd	2.158a	12.1ab	11.4abc
Xh19	1.550bc	2.206a	11.6ab	12.3ab
Xh20	1.603b	2.224a	11.9ab	12.6a
Xy21	1.688a	2.224a	10.9bc	11.2abc
Xy24	1.688a	2.242a	9.4cde	8.6de

Table 3. N absorption and use efficiency of different oilseed rape varieties

3.3 The differences of N efficiency in different varieties and its analysis

Results of Table 2 and Table 3 showed that, the differences of N efficiency, N absorption efficiency and N use efficiency between varieties were significant under no N application condition; and the differences of N efficiency and N use efficiency between varieties were significant under N application condition; while there were no differences of N absorption efficiency between varieties. Grain yield of high N efficiency variety is higher than average yield of 16 varieties under no N application condition; otherwise it was low N efficiency variety. D-value of grain yield of high N response variety between no N application and N application conditions is higher than average D-value of 16 varieties; otherwise it is low N response variety. The 16 varieties can be divided into 4 genotypes according to the definitions (Table4): (1) N high efficiency – N high response; yield was higher under no N application condition; yield was increased significantly under N application condition (see varieties Xy1,Xy16,Xy17,Xh19,Xh20 and Xy21); (2) N low efficiency – N low response; yield was lower under no N application condition; yield increase was not distinct under N application condition; (see varieties Xy6, Xy8 and Xy9); (3) N high efficiency – N low response, yield was higher under no N application condition; but the yield increase was not distinct under N application condition (see varieties Xy7,Xy12,Xy14,Xy15 and Xy24); (4) N low efficiency – N high response; yield was lower under no N application condition, but the yield was increased significantly under N application condition (see varieties Xy11 and Xy13).

Type	NHE-NHR	NLE-NLR	NHE-NLR	NLE-NHR
Variety	Xy1 Xy16 Xy17 Xy19 Xy20 Xy21	Xy6 Xy8 Xy9	Xy7 Xy12 Xy14 Xy15 Xy24	Xy11 Xy13

Table 4. N efficiency and response type of oilseed rape

Results of Table5 showed that, variation coefficient of grain yield were almost the same regardless of N application level. Variation coefficient of N use efficiency was higher than variation coefficient of N absorption amount under the two N application level conditions. It was suggested that, the contribution of N use efficiency to N efficiency was higher than N

N Levels		Mean	S	CV(%)
No N application	Grain yield (kg/hm²)	1584	323	20.4
	N absorption (g/Plant)	1.499	0.142	9.5
	N use efficiency (g/g)	10.5	1.7	16.2
N application	Grain yield (kg/hm²)	2116	424	20.0
	N absorption (g/Plant)	2.194	0.031	1.4
	N use efficiency (g/g)	9.7	1.9	19.5
Subtraction of both		533	277	52.1

Table 5. Variation of Grain yield, N absorption and use efficiency of different oilseed rape varieties

absorption efficiency in oilseed rape under the field condition, regardless of N application level. The differences of variation coefficient between N absorption amount and N use efficiency were smaller under no N application condition, compared with N application condition. It was suggested that variation coefficient of N use efficiency was declined, but variation coefficient of N absorption amount was increased under N stress condition. This indicates that, the contribution of N absorption efficiency to N efficiency was increased with N application level, but the contribution of N use efficiency to N efficiency was decreased.

3.4 Biomass and N absorption of the two oilseed rape varieties

Table 6 results showed that there is no significant difference of N content in plant and grain between X-36 and X-50, in addition, biomass, harvesting indexes and total N amount of X-36 are almost the same with X-50. However, grain yield and N harvesting indexes of X-36 are higher than X-50 significantly. This indicates that the reasons for grain yield of X-36 being higher than X-50 was not greater N absorption and harvesting indexes but rather higher N redistribution amount and higher N harvesting index.

Variety	Biomass (g⊙plant¹)	Grain yield (g⊙plant¹)	N concentration of plant (mg⊙g⁻¹DW)	N concentration of grain (mg⊙g⁻¹DW)	Total N in plant (g⊙plant¹)	Total N in grain (g⊙plant¹)	Harvesting index	N harvesting index
X-36	119.8±9.5	31.0±2.6*	27.0±0.9	44.2±1.8	3.22±0.16	1.37±0.07	0.26±0.024	0.43±0.029*
X-50	111.4±8.7	25.7±2.9	28.1±1.8	45.5±1.7	3.12±0.23	1.17±0.14	0.23±0.018	0.37±0.029

*Indicated that differences between varieties in t-test are significant at p<0.05 level. The same as below.

Table 6. Physiological parameters of two oilseed rape varieties

3.5 Distribution characters of N absorbed at different growth stages

Fig.1 results showed that N absorbed at seedling and stem elongation stages are mainly distributed into leaves, average proportions of transferred-N/total-N in two varieties are accounting for 83.5% and 66.3% respectively. N absorbed at flowering stage is mainly distributed into leaves and stem, proportions are accounting for 42.8% and 36.3% respectively. While 42% (average values of two cultivars) of N absorbed at siliquing stage is directly distributed into silique. Proportion of N distributed into leaves is declined on a large scale with senescence of plant, proportion of N distributed into stem and root are declined also, however proportion of N distributed into silique and grain are increased at the same time. Proportion of N absorbed at seedling stage is redistributed from vegetative organs to grain in X-50 is the lowest, but still 43% of N is distributed into grain at harvesting stage.

In addition, proportion of N that was distributed into leaves at flowering stage in X-36 and X-50 are 47.3%, 40.7% respectively, and siliquing stages are 30.6%, 25.8% respectively. Proportion of N that was distributed into grain at harvesting stage in X-36 are higher than X-50, and proportion differences between the two varieties of N absorbed at seedling and flowering stages are significant (Fig.1 and Table7). Higher proportion of N distributed into leaves is a benefit for normal functions of leaves development, higher proportion of N distributed into grain is benefit for N harvesting indexes improvement.

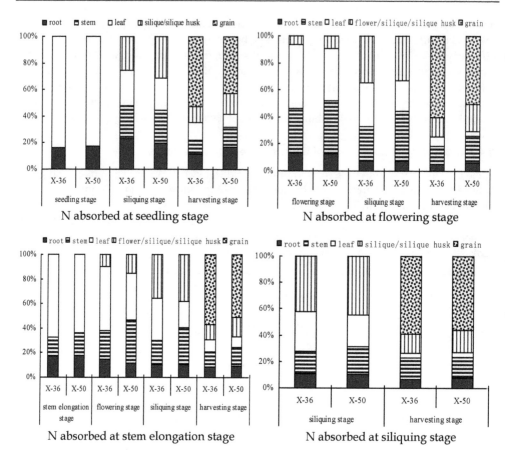

Fig. 1. Distribution proportion of N which absorbed at different growth stages in oilseed rape (vertical axis is stand for N distribution proportion among plant organs, lateral axis is stand for different growth stages)

Physiological indexes	Variety	N absorbed at seedling stage	N absorbed at stem elongation stage	N absorbed at flowering stage	N absorbed at siliquing stage	Sum
Amount of N transferred (mg)	X-36	231.2*	325.5	275.5*	88.9*	921.1
	X-50	175.1	326.0	160.4	75.0	736.5
Proportion of N transferred (%)	X-36	38.2*	46.8	49.6*	32.8	
	X-50	30.6	41.8	32.8	30.6	
Proportion of transferred-N/ grain-N (%)	X-36	16.9	23.8	20.1*	6.5	67.2
	X-50	15.0	27.9	13.7	6.4	62.9

Table 7. Translocation to grain of N which absorbed at different stages

3.6 Distribution proportion of N absorbed at different growth stages

Table7 results showed that distribution proportion of N absorbed at stem elongation and flowering stages are higher than other stages, average proportion at stem elongation and flowering stages are 44.3% , 41.2% respectively, and 34.4% , 31.7% respectively at seedling and siliquing stages. Distribution amount of N absorbed at stem elongation stage is the highest, average value is 325.8 mg, following are N absorbed at flowering and seedling stages, average values are 218.0 mg and 203.2 mg respectively, the lowest is N absorbed at siliquing stage, average value is 82.0 mg.

Average proportion of transferred-N/grain-N in two varieties is 65.1%. Distribution proportion of N absorbed at stem elongation stage is the highest, accounting for 25.8%, following are N absorbed at flowering and seedling stages, accounting for 16.9 % and 15.9% respectively, the lowest is N absorbed at siliquing stage, accounting for 6.4%. Table3 results also showed that distribution amount of N absorbed at seedling, flowering and harvesting stages in X-36 are higher than X-50 significantly, distribution proportion of N absorbed at seedling and flowering stages in X-36 are higher than X-50 significantly.

N absorbed at seedling and stem elongation stages that was distributed into reproductive organs, it is real N that redistributed from vegetative organs to reproductive organs. However, part of N absorbed at flowering and siliquing stages was directly transferred into reproductive organs, therefore total N in reproductive organs is not only because of N redistribution, but also because of root absorption. So, differences amount of N, which was absorbed at flowering and siliquing stages, between absorption and redistribution can not be distinguished in this study, but the main results can not be affected by these influences significantly.

3.7 Loss amount and proportion of N at harvesting stage

Table 8 results showed that loss proportion and amount of N absorbed at seedling stage are the largest, average values of the two varieties are 24.0% and 141.6 mg, the next is N absorbed at stem elongation and flowering stages, average proportion values are 10.5% and 11.7%, average amount values are 79.2 mg and 43.2 mg respectively, the smallest loss are N absorbed at siliquing stage, average values are 7.3% and 16.2 mg. Table 4 also showed that N loss proportion and amount of X-36 are lower than X-50, but there are no significant different. Loss amount of N absorbed at stem elongation and flowering stages in X-36 (63.0 and 35.0 mg) are lower than X-50 (95.3 and 51.4 mg) significantly. Total loss amount of N absorbed at the four growth stages in X-36 (252.3 mg) is 18.1% lower than X-50 (307.9 mg).

Physiological indexes	Variety	N absorbed at seedling stage	N absorbed at stem elongation stage	N absorbed at flowering stage	N absorbed at siliquing stage	Sum
N Loss amount plant^{-1} (mg)	X-36	136.4	63.0	35.0	17.9	252.3
	X-50	146.7	95.3*	51.4*	14.5	307.9
N Loss proportion (%)	X-36	22.8	9.0	9.6	8.5	
	X-50	25.1	12.0	13.8	6.1	

Table 8. Loss amount and proportions of N at harvesting stage (N absorbed at different stages)

Components of N can not be reflected completely in Table8. This experiment was carried out in glass room, so possibility of rain wash can be ignored. Old and dropped leaves were carefully collected during the experiment, so N loss through senescence and leaf dropped can be neglected. N loss through water exudation and dew was little also. Therefore, only N loss through N volatilization from the aboveground organs and root secretion can not be calculated in this study.

4. Discussion

4.1 N efficiency in different oilseed rape varieties

Environmental contamination was reduced by less N fertilizer application, so, breeding and spreading new crop varieties with high N efficiency was an interesting topic recently (Xu et al., 2006; Pei et al., 2007 and Wang et al., 2003). Many studies and reports were involved in N agriculture efficiency, N physiological efficiency and N absorption efficiency. N physiological efficiency and N absorption efficiency were parts of the N agriculture efficiency, there were relationships between the three efficiency indexes. Wheat breeders identified selection standards based on physiological indexes, such as high yield and high protein content (Monaghan et al., 2001). Results showed that, the differences of N efficiency in 16 oilseed rape varieties were significant, regardless of N application level. The results obtained in the present study are consistent with previous published results (Beauchamp et al., 1976; Pollmer et al., 1979; Balko et al., 1980a; Reed et al., 1980; Russell, 1984; Moll et al., 1987; Landbeck, 1995; Bertin et al., 2000). However, few oilseed rape varieties were used in these studies, more varieties were needed to be selected and estimated in the future. 16 oilseed rape varieties were used as plant materials in this experiment, N high efficiency types were better plant materials and can tolerate low N stress. These materials have high economic benefit under low N condition; N high response types were better plant materials and can tolerate high N stress, these materials have high economic benefit under high N condition. It can be seen that, the type of "N high efficiency – N high response" varieties was the ideal variety.

4.2 N efficiency of oilseed rape under different N application level

Breeding procedure was usually conducted under N application conditions recently. So, the varieties have high yield under N application condition, it was only for N application condition. It was necessary to select high N efficiency varieties under the low N stress condition, select the varieties have high yield under no N application level (Cao et al., 2010). The correlation of various agronomic traits (grain protein, yield and its component) were different with the changes of N application level (Balko et al., 1980b; Fonzo et al., 1982; Rizzi et al., 1993; Bertin et al., 2000). Therefore, the response differences of varieties under different N application level conditions were obscured; these problems are needed to be further studied in the future. Möllers et al., (2000) studies suggested that, the differences of N efficiency between different N application levels and between varieties were big, observed significant interactions between genotype and N level suggested that the high yielding genotypes in high N supply were not necessarily high yielding in the low N supply. This study has the same results, the varieties have the same yield under N application condition, but under no N application condition, the differences of yield between varieties were big. The results of the present study is different from the results of

Grami et al. (1977) which showed that a direct relationship existed between N uptake and seed N content; they concluded that high seed N content was good for N uptake and translocation efficiency. The above results suggested that, both of N application condition and no N application conditions should be considered during the breeding and selecting procedure of crop (Liu et al., 1999; Lafitte et al., 1994).

4.3 The contribution of N absorption efficiency and N use efficiency to N efficiency

Many studies were carried out on the in contribution of N absorption efficiency and N use efficiency to N efficiency, but the results were different (Zou et al., 2009). High seed protein content will cause low grain yield, this agrees with former results which observed negative relationship between protein content of seed and grain yield (Dudley et al., 1977; Simmonds, et al., 1995). Wheat breeders have reported selection standards combined with high yield and high protein content (Monaghan et al., 2001). Some results showed that N use efficiency was the main reason for changes of N efficiency under low N condition, and the main reason was N absorption efficiency under high N condition (Moll et al., 1982). The effect of N absorption efficiency on N efficiency was higher than N use efficiency significantly under low N condition, and the effect of N absorption efficiency and N use efficiency on N efficiency were almost the same under high N condition (Mi et al., 1998); the absorption efficiency was the main reason for changes of N efficiency regardless of N application level (Liu et al., 2002); N absorption efficiency and N use efficiency of oilseed rape were studied by Yan and Thurling (1987a), there were differences of N use efficiency between varieties under high N condition, and there were differences of N absorption efficiency between varieties under low N condition. The results of this study showed that, the changes of N use efficiency was the main reason of the changes of N efficiency under low and high N conditions, but Yan and Thurling (1987b) have different results. This study (Table 5) also showed that, variation coefficient of N absorption efficiency was increased under low N condition, while the variation coefficient of N use efficiency declined, it was suggested that, the contribution of N absorption efficiency to N efficiency was increased under low N condition. However, the results were only for the varieties which were used in this study, and more varieties are necessary to be used in the future.

4.4 N redistribution characteristics and relation to N efficiency

Liu, (1987) reported that N content in oilseed rape tissues at different growth stages are from 4.5% to 1.2%, N content is higher during earlier growth stages, and lower during later growth stages, N distribution and transfer in tissues can reflect plant metabolisms situation and growth center changes. The same results are showed in this study (Fig.1), N absorbed at seedling and stem elongation stages are mainly distributed into leaves, and distributed amount of N absorbed at flowering stage is lower than the former stages, 42% of N absorbed at siliquing stage was directly distributed into silique. N redistribution is happen in every part of plant tissues at different growth stages, especially after flowering stage, large amount of N are redistributed from vegetative organs to reproductive organs, these results are reported in many kinds of crops, such as wheat (Andersson et al., 2005; Palta et al., 1995a; Palta et al., 1995b; Dalling et al., 1985; Barbottin et al., 2005; Pierre et al., 2003), rice (He et al., 2002), maize (Pommel et al., 2006), pea (Séverine et al., 2005), and oilseed rape (Rossato et al., 2002; Malagoli et al., 2005). Malagoli et al., (2005) suggested that N requirements of seed filling are mainly satisfied by N mobilized from vegetative parts (about 73% of total N in pods). Nearly all that had been remobilized from vegetative tissues were finally recovered in mature pods (Rossato et al., 2002). The results of maize (Liu et al., 2002) and wheat (Li et al., 2006) studies showed that contribution of N redistributed from vegetative organs to grain was about 50-95%, and depended on crop growth conditions, variety and N application. Table3 results also show that contribution of N redistributed from vegetative organs to grain was 65.1%, this confirmed the results of the other studies. However, a comparison of the amount of N redistribution and the proportion of N absorbed at different growth and development stages has not been done (Rossato et al., 2002; Malagoli et al., 2005; Liu et al., 2002; Li et al., 2006). Preliminary results are shown in this study (Table7), transferred proportion of N absorbed at stem elongation stage was the highest, accounting for 25.8%, followed by N absorbed at flowering and seedling stages, accounting for 16.9% and 15.9% respectively, the lowest was absorbed at siliquing stage, accounting for 6.4%.

High efficiency of N redistribution during later growth stages is physiological mechanisms for crop adapting to environmental changes for a longtime. It is important for crop to alleviate N deficiency in plant tissues, improve crop production and N efficiency, and protect environment ecology (Zhang et al., 2010). However, studies on the N redistribution and loss of oilseed rape are few recently; comparatively studies on the differences of N distribution between growth stages of oilseed rape are fewer (Zhang et al., 2010). This paper system studied on amount and proportion of N that was absorbed at different growth stages, and N distribution, loss amount during later growth stages. Results (Table8) showed that amount and proportion of N absorbed at stem elongation stage that is redistributed into grain of the two varieties was the highest, and loss proportion of N is lower than the other stages also. Loss amount and proportion of N absorbed at seedling stage is the highest, and distribution amount and proportion value is moderate. It can be concluded that fertilizer application at stem elongation stage is good for improving the N application efficiency.

5. Acknowledgments

This study was supported by the National Natural Science Foundation of China (Grant No. 31071851, 31101596 and 30971860), Talent Introduction Strategy of Hunan Agricultural University (11YJ21) and FuRong Scholar Program of Hunan Province, P.R. China.

6. References

Andersson, A., E. Johansson, and P. Oscarson. 2005. Nitrogen redistribution from the roots in post-anthesis plants of spring wheat. *Plant and Soil* 269: 321-332.

Balko, L.G., Russell W.A., 1980a. Effects of rates of nitrogen fertilizer on maize inbred lines and hybrid progeny. I. Prediction of yield response. Maydica., 25: 65-79.

Balko, L.G., Russell W.A., 1980b. Effects of rates of nitrogen fertilizer on maize inbred lines and hybrid progeny. II. Correlations among agronomic traits. Maydica., 25: 81-94.

Barbottin, A., C. Lecomte, C. Bouchard, and M. H. Jeuffroy. 2005. Nitrogen remobilization during grain filling in wheat: genotypic and environmental effects. *Crop Science* 45: 1141-1150.

Beauchamp, E.G., Kannenberg L.W., Hunter R.B., 1976. Nitrogen accumulation and translocation in corn genotypes following silking. Agronomy Journal., 68:418-422.

Bertin, P. and Gallais, A., 2000. Genetic variation for nitrogen use efficiency in a set of recombinant maize inbred lines. Agro., physiological results. Maydica., 45: 53-63.

Brenna, T. J., T. N. Corso, H. J. Tobias, and R. J. Caimi. 1998. High-precision continuous-flow isotope ratio mass spectrometry. *Mass Spectrometry Reviews* 16(5): 227 – 258.

Cao, L.Q., Wu, X.M., Li, Y.J. 2010. Relationship between genotypic differences of rapeseed (*Brassica napus* L.) nitrogen uptake efficiency and economic characteristics. *Chin. J. Oil Crop Sci.* 32(2): 270-278.

Christian, M., Maria, K., Bettina, K., Ariane, O. and Heiko, C. B., 1999. Genotypic variation for nitrogen efficiency in winter rapeseed cultivars. "New Horizons for an old crop", Proceedings of the 10th international rapeseed congress., Canberra, Australia.

Dalling, M. J. 1985. The physiological basis of nitrogen redistribution during grain filling in cereals. In: *Harpen JE, Schrader LE, Howell RW (Eds.)*. Exploitation of physiological and genetic variability to enhance crop productivity. p:55-71.

Dimmock, J., and M. J. Gooding. 2002. The influence of foliar diseases, and their control by fungicides, on the protein concentration in wheat grain: a review. *Journal of Agricultural Science* 138: 349-366.

Dong, G. C., Y. L. Wang, J. Zhou, B. Zhang, C. S. Zhang, Y. F. Zhang, L. X. Yang, and J. H. Huang. 2009. Difference of nitrogen accumulation and translocation in conventional indica rice cultivars with different nitrogen use efficiency for grain output. *Acta Agron. Sin.* 35(1): 149–155.

Dreccer, M. F., Schapendonk, A. H. C. M., Slafer G. A. and Rabbinge, R., 2000. Comparative response of wheat and oilseed rape to nitrogen supply: absorption and utilization efficiency of radiation and nitrogen during the reproductive stages determining yield. Plant and Soil., 220: 189–205.

Dudley, J.W., Lambert. R.J. and Roche, D.E. I.A., 1977. Genetic analysis of crosses among corn strains divergently selected for percent oil and protein. Crop Sci., 17: 111-117.

Fonzo, D. N., Motto M., Maggiore, T., Sabatino, R., Salamini, F., 1982. N uptake, translocation and relationships among N related traits in maize as affected by genotype. Agronomie., 2: 789-796.

Grami, B. and La Croix, L.J., 1977. Cultivar variation in total nitrogen uptake in rape. Can. J. Plant Sci., 57: 619-624.

Hardy, R. W. F., Holsten, R. D., Jackson E. K. and Burns, R. C., 1968. The Acetylene-Ethylene Assay for N2 Fixation: Laboratory and Field Evaluation. Plant Physiology., 43:1185-1207.

He, P., W. Zhou, and J. Y. Jin. 2002. Effect of nitrogen application on redistribution and transformation of photosynthesized ^{14}C during grain formation in two maize cultivars with different senescence appearance. *Journal of Plant Nutrition* 25(11):2443-2456.

Kamprath, E. J., R. H. Moll, and M. Rodrigues. 1982. Effects of nitrogen fertilization and recurrent selection on the performance of hybrid populations of corn. *Agronomy Journal* 74: 955-958.

Kessel, B. and Becker, H.C., 1999. Genetic variation of nitrogen-efficiency in field experiments with oilseed rape (Brassica napus L.). "New Horizons for an old crop", Proceedings of the 10th international rapeseed congress., Canberra, Australia.

Landbeck, M.V., 1995. Untersuchungen zur genetischen verbesserung der anbaueigung von koÈrnermais unter produktionsbedingungen mit verringerter sticksoffversorgung. Dissertation, University of Hohenheim.

Lafitte, H. R., Edmeades, G. O., 1994. Improvement for tolerance to low soil nitrogen in tropical maize Selection criteria field. Crop Research., 39: 1-14.

Li, S. W., H. D. Wen, Y. Z. Zhou, Y. M. Li, and K. Xiao. 2006. Characterization of nitrogen uptake and dry matter production in wheat varieties with different N efficiency. *Scientia Agricultura Sinica* 39 (10): 1992-2000.

Liu, H. L. 1987. Practical Cultivation in Oilseed Rape. Shanghai: *Shanghai Scientific and Technical Publishers* pp: 128-143.

Liu, J. M., Mi, G. H., Zhang, F S., 1999. Difference in nitrogen efficiency among maize genotypes. J. Agric. Tural. Biotech., 7（3） : 248-254.

Liu, J. A., Mi, G H., Chen, F. J., Zhang, F. S., 2002. Genotype differences on nitrogen use efficiency among maize hybrids. Plant Nutr. Fert . Sci., 8（3）:276-281.

Liu, Q., Song, H.X., Rong, X.M. 2009. Studies on oilseed yield and nitrogen efficiency in different cultivars of oilseed rape (*Brassica napus* L.). *Plant Nutri. Ferti. Sci.* 15(4): 898-903.

Mahmoud Kamh, Franz Wiesler, Abdullah Ulas, Walter J. Horst, 2005. Root growth and N-uptake activity of oilseed rape (Brassica napus L.) cultivars differing in nitrogen efficiency. Journal of Plant Nutrition and Soil Science., 168:130 – 137.

Malagoli. P., P. Laine, L. Rossato, and A. Ourry. 2005. Dynamics of nitrogen uptake and mobilization in field-grown winter oilseed rape (*Brassica napus L.*) from stem extension to harvest. *Annals of Botany* 95: 853-861.

Mi, G. H., Liu, J. A., Zhang, F. S., 1998. Analysis on nitrogen agronomic efficiency and its components of maize hybrids. J. of China Agric. Univ., 3: 97- 104.

Moll, R. H., Kamprath, E. J. and Jackson, W. A., 1982. Analysis and interpretation of factors which contribution efficiency of nitrogen utilization. Agron. J., 74: 562-568.

Moll, R. H., Kamprath, E.J., Jackson W.A., 1987. Development of nitrogen efficient hybrids of maize. Crop Science., 27:181-186.

Möllers, C., Kessel, B., Kahlmeyer, M., 2000. Ossenkop A and Becker H C. Untersuchungen zur genotypischen Variabilität der Stickstoff-effizienz bei Winterraps. In: Möllers C. (Hrsg.) Stickstoff-effizienz Landwirtschaftlicher Kulturpflanzen. Erich Schmidt Verlag.

Monaghan, J.M., Snape, J.W.,Chojecki. A.J.S, and Kettlewell, P.S., 2001. The use of grain protein deviation for identifying wheat cultivars with high grain protein concentration and yield. Euphytica., 122: 309-317.

Norusis, M. J., 1990. SPSS advanced statistics user's guide. S Inc SPSS Chicago.

Palta, J. A., and I. R. P. Fillery. 1995a. N application increases pre-anthesis contribution of dry matter to grain yield in wheat grown on a duplex soil. *Australian Journal of Agricultural and Research* 46: 507–518.

Palta, J. A., and I. R. P. Fillery. 1995b. N application enhances remobilisation and reduces losses of pre-anthesis N in wheat grown on an Duplex soil. *Australian Journal of Agricultural and Research* 46: 519–531.

Palta, J. A., T. Kobata, N. C. Turner, and R. Fillery. 1994. Remobilization of carbon and nitrogen in wheat as influenced by post-anthesis water deficits *Crop science* 34: 118-124.

Pei, X. X., Wang, J. A., Dang, J. Y., Zhang, D. Y., 2007. An approach to the screening index for low nitrogen tolerant wheat genotype. Plant Nutrition and Fertilizer Science., 13(1): 93 – 98.

Peoples, M. B., and M. J. Dalling. 1988. The Interplay between Protelysis and Amino Acid Metabolism during Senescence and Nitrogen Reallocation. In: Nooden L D, Leopold A C eds. Senescence Aging in Plant. *San Diego: Academic Press* pp: 181–217.

Pierre, M., R. P. John, D. J. Peter, and T. Eugène. 2003. Modeling grain nitrogen accumulation and protein composition to understand the sink/source regulations of nitrogen remobilization for wheat. *Plant Physiology* 133:1959-1967.

Pollmer, W. G., Eberhard, D., Klein, D., Dhillon, B.S., 1979. Genetic control of nitrogen uptake and translocation in maize. Crop Science., 19:82-85.

Pommel, B., A. Gallais, I. Coque, B. Hirel, J. L. Prioul, B. Andrieu, and M. Floriot. 2006. Carbon and nitrogen allocation and grain filling in three maize hybrids differing in leaf senescence. *European Journal of Agronomy* 24: 203-211.

Przulj, N., and V. Momcilovic. 2001. Genetic variation for dry matter and nitrogen accumulation and translocation in two-rowed spring barley. II. Nitrogen translocation. *European Journal of Agronomy* 15: 255-265.

Rathke, G.W. and Behrens, T., 2006. Integrated nitrogen management strategies to improve seed yield, oil content and nitrogen efficiency of winter oilseed rape (Brassica napus L.): A review. Agriculture, Ecosystem Enviroment., 117: 80-108.

Reed, A. J., Below, F. E., Hageman, R. H., 1980. Grain protein accumulation and the relationship between leaf nitrate reductase and protease activities during grain development in maize. I. Variation between genotypes. Plant Physiology., 66:164-170.

Rizzi, E., Balconi, C., Nembrini, L., Stefanini, F.M., Coppolino, F., Motto, M., 1993. Genetic variation and relationships among Nrelated traits in maize. Maydica., 38: 23-30.

Rossato, L., L. C. Dantec, P. Laine, and A. Ourry. 2002. Nitrogen storage and remobilization in *Brassica napus* L. during the growth cycle: identification, characterization and immunolocalization of a putative taproot storage glycoprotein. *Journal of Experimental Botany* 53: 265-275.

Rossato, L., P. Laine, and A. Ourry. 2001. Nitrogen storage and remobilization in *Brassica napus* L. during the growth cycle: nitrogen fluxes within the plant and changes in soluble protein patterns. *Journal of Experimental Botany* 52(361): 1655-1663.

Russell, W. A., 1984. Further studies on the response of maize inbred lines to N fertilizer. Maydica., 29: 141-150.

Sattelmacher, B., Horst, W.J. and Becker, H.C., 1994. Factors that contribute to genetic variation for nutrient efficiency of crop plants. Z. Pflanzenernähr Bodenk., 157, 215-224.

Schjoerring, J. K., Bock, J. G. H., Gammelvind, L., Jensen, C.R. and Mogensen, V.O., 1995. Nitrogen incorporation and remobilization in different shoot components of field-grown winter oilseed rape (Brassica napus L.) as affected by rate of nitrogen application and irrigation. Plant and Soil., 177: 255-264.

Seiffert, B., Zhou, Z., Wallbraun, M., Getrud, L. and Christian, M., 2004. Expression of a bacterial asparagines synthetase gene in oilseed rape (Brassica napus L.) and its effect on traits related to nitrogen efficiency. Physiologia Plantarum., 121(4): 656-665.

Séverine, S., M. J. Nathalie, J. Christian, B. Judith, and S. Christophe. 2005. Dynamics of exogenous nitrogen partitioning and nitrogen remobilization from vegetative organs in pea revealed by ^{15}N in vivo labeling throughout. *Plant Physiology* 137: 1463-1473.

Simmonds, N.W., 1995. The relation between yield and protein in cereal grain. J. Sci. Food Agric., 67: 309-315.

Thomas, K., H. Bertrand, H. Emmanuel, D. Frédéric, and L. G. Jacques. 2007. In winter wheat (*Triticum aestivum* L.), post-anthesis nitrogen uptake and remobilization to the grain correlates with agronomic traits and nitrogen physiological markers. *Field Crops Res.* 102: 22–32.

Wang, H., T. N. McCaig, R. M. Depauw, F. R. Clerke, and J. M. Clerke. 2003. Physiological characteristics of recent Canada western red spring wheat cultivars: Components of grain nitrogen yield. *Can. J. Plant Sci.* 83(4): 699–707.

Wang, Y., Mi, G. H., Chen, F. J., Zhang, F. S., 2003. Genotypic differences in nitrogen uptake by maize inbred lines its relation to root morphology. Acta Ecol . Sin., 23 (2): 297-302.

Wiesler, F., Behrens, T., and Horst, W. J., 2001a. The role of nitrogen-efficient cultivars in sustainable agriculture. The Scientific World Journal., 1(S1):61-69.

Wiesler, F., Behrens, T., and Horst, W. J., 2001b. Nitrogen efficiency of contrasting rape ideotypes. In: Horst W J et al. (eds.) Plant Nutrition - Food Security and Sustainability of Agro-Ecosystems, Kluwer Academic Publishers., 60-61.

Xu, X. Y., Zhang, M. M., Zhai, B. N,, LI, S. X., Zhang, X. C., Wang, Z. H., 2006. Genotypic variation in nitrogen use efficiency in summer maize. Plant Nutrition and Fertilizer Science., 12(4) :495-499.

Yan, S. K. and Thurling, N., 1987a. Variation in nitrogen response among spring rape (*Brassica natus* L.) cultivars and its relationship to nitrogen uptake and utilization. Field Crops Research., 16: 139-155.

Yan, S. K. and Thurling, N., 1987b. Genetic variation in nitrogen uptake and utilization in spring rape (*Brassica natus* L.) and its exploitation through selection. Plant Breeding., 98: 330-338.

Zhang, Y. L., and Z. W. Yu. 2008. Effects of irrigation amount on nitrogen uptake, distribution, use, and grain yield and quality in wheat. *Acta Agron. Sin.* 34(5): 870–878.

Zhang, Y. H., J. Wu, Y. L. Zhang, D. S. Wang, and Q. R. Shen. 2006. Genotypic variation of nitrogen accumulation and translocation in japonica rice (*Oryza sativa* L.)cultivars with different height. *J. Nanjing Agric. Univ.* 29(2): 71–74.

Zhang, Y. H., Y. L. Zhang, Q. W. Huang, Y. C. Xu, Q. R. Shen. 2006. Effects of different nitrogen application rates on grain yields and nitrogen uptake and utilization by different rice cultivars. *Plant Nutr. Fert. Sci.* 12(5): 616–621.

Zhang, Z. H., H. X. Song, Q. Liu, X. M. Rong, C. Y. Guan, J. W. Peng, G. X. Xie, and Y. P. Zhang. 2010. Studies on differences of nitrogen efficiency and root characteristics of oilseed rape cultivars in relation to nitrogen fertilization. *Journal of Plant Nutrition* 33:1448-1459.

Zhang, Z. H., H. X. Song, Q. Liu, X. M. Rong, J. W. Peng, G. X. Xie, Y. P. Zhang, C. Y. Guan. 2010. Nitrogen redistribution characteristics of oilseed rape varieties with different nitrogen-use efficiencies during later growth period. *Communications in Soil Science and Plant Analysis* 41: 1693-1706.

Zou, J., Lu, J.W., Chen, F. 2009. Effect of Nitrogen, Phosphorus, Potassium, and Boron Fertilizers on Yield and Profit of Rape seed (*Brassica napus* L.) in the Yangze River Basin. *Acta Agronomica Sinica.* 35(1): 87–92*.

5

Oilseed Pests

Masumeh Ziaee

Islamic Azad University, Shoushtar Branch,
Iran

1. Introduction

Oilseeds occupy an important position in agriculture and industrial economy. So, management of pest problems and using possible control techniques could increase the quality and quantity of the products.

A large variety of pests damage oilseeds and cause significant losses in the farms or storages. Among which insect pests are a significant factor in the economics of oilseed production. There are also several vertebrate pests such as rats, slugs and birds.

An extensive review of oilseed pest has been conducted by numerous national research programs in Australia, Canada, Germany, India, United Kingdom, United State and etc.

Pests of oilseeds can be classified according to the different factors such as taxonomic grouping, feeding habitat, distribution. Within each groups, the pest status may vary with country, year and season depending on species, variety, cropping system used. Some of the pests may attack a certain type of oilseed; the others are generalists and can feed on a variety of crops. Pests with different feeding habitat could have either chewing or sucking mouthparts. Pests with chewing mouthparts eat parts of the crops such as flowers, foliage, stems, roots or buds. They may only eat portions of leaves, leave holes in the leaves (flea beetles), cause bud abscission, reducing yield of the pods and leaf area, earlier leaf abscission, delayed flowering (weevils) or bore into stems and roots (borers).

Pests with sucking mouthparts such as aphids, bugs, thrips, whiteflies, mites, and jassids usually cause the plant to discolor or twist and curl. The plant may discolor from tiny yellow speckles (spider mites), larger darkened spots (plant bugs), reducing the canopy area of the plant, and therefore its photo-synthetic capacity, coatings of black sooty mold growing on honeydew deposits or transmitting viruses (aphids and whiteflies).

Also, several pests may attack oilseed crops occasionally or their populations don't exceed threshold levels.

Protection of the crops from pest's infestations and keeping the pests under proper control has become in consideration due to the importance of the crops. Pest management is done by different methods such as cultural control, biological control, physical control, host plant resistance and assessing the economic thresholds to determine the need to apply pesticides (chemical control). By doing one of these methods, farmers can protect their farms and agricultural produce. However, integrating two or more control techniques to manage one or more species of the same group of pests as an IPM (Integrated Pest Management) program prevent pests from causing significant losses, encouraging natural enemies, saving money while producing a high quality product, enhance the agricultural productivity and usually has the highest probability of cost effectiveness (Fig. 1).

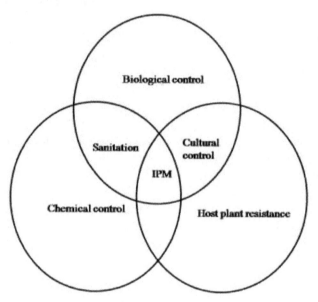

Fig. 1. IPM diagram

In this chapter, the order of presentation is firstly related to the oilseed crops. Although the importance of crops differ from region to region but some most prominent crops are cotton seeds, groundnuts, mustard, rapeseed, safflower, sesame, soybean, sunflower, linseed and castor. Secondly, within each crop, the pests are presented by taxonomic groups; attempt to help you identify the common or special pests. Subsequently, the life cycle, monitoring, economic importance, action thresholds of the pests and methods for keeping pests under control are given. In summary the best management program as IPM will be introduced.

The format for a full entry is:
Oilseed crop: Common and scientific name of oilseed crop.
Pests: arranged by taxonomic order.
Identification: Size of insect and distinctive coloration.
Life cycle: the pest's life cycle, habitat, habits, damage stage and detection of the pests.
Monitoring: monitor and detect significant populations of the species.
Economic importance and action thresholds: number of pests, density, etc. at which a pest becomes an economic threat and control is essential. The action level also depends on costs and the economic benefits of treatment.
Control: different methods can be used to control pests such as cultural control, biological control, physical control, host plant resistance and chemical control.
-    Cultural control: Cultural control exploits the factors related to reduce the prevalence of unwanted pests. Crop rotation, intercropping, sanitation, early or late planting, trap

crops, fertilization, and cold and heat treatments are strategies that may be highly effective for controlling some pests.
- Biological control: biological control is the use of one organism to control a particular pest. Biological control agents include predators, parasitoids, and pathogens.
- Physical control: physical control is the use of some physical components of the environment, such as temperature, humidity, light, electromagnetic radiation (microwaves, infrared, and radiofrequencies), mechanical shock, and pneumatic control (blowing or vacuuming tools) to the detriment of pests.
- Host plant resistance: Planting resistant varieties can prevent or decrease pest damages and their injuries to the crop.
- Chemical control: The most common method of pest control is the use of chemicals that either kill pests or inhibit their development. A well chosen insecticide provides the desired level of control, while minimizing negative side effects such as enhancing the development of resistance, hazards effects on non-target species and beneficial organisms, environmental contamination and flaring secondary pests.

IPM program: IPM attempts to find the optimum combination of control tactics including cultural, biological, physical, and chemical control that will reduce pest population below the economic threshold. IPM is a safer, usually less costly and reduced-risk option for effective pest management.

The life cycle of the pests, damage and economic thresholds depend on many factors (crop stage, crop age, socio-economic, climatic conditions and etc.) and cannot be adopted without consideration to local conditions. So, descriptions of each pest are gathered from available literatures and researches depending on the importance of the crop and invasive pests.

Conclusion: at the end of this chapter we could identify the pests and beneficial species community on the oilseed crop, distribution of them on the crop and have information about the infestation. After considering safety, effectiveness, environmental effects and cost of each method; we choose the best method for protecting the crops and minimizing pest's problems.

The oilseed crops listed in this chapter are 1) cottonseed, 2) groundnuts, 3) soybean, 4) mustard, 5) sunflower, and 6) safflower, respectively.

2. Cottonseeds

There are five commercially grown species of cotton:

- *Gossypium hirsutum* L.: known as upland cotton or Mexican cotton, native to Central America and Mexico (90% of world production).
- *Gossypium barbadense* L.: known as Creole, Egyptian, South American, Pima or Sea Island cotton, native to tropical South America (8% of world production).
- *Gossypium arboreum* L.: known as tree cotton, native to India and Pakistan (less than 2%)
- *Gossypium herbaceum* L.: known as Levant cotton, native to semi-arid regions of sub Saharan Africa and the Arabian Peninsula (less than 2%).
- Organic cotton: grown in subtropical countries such as America and India (growing at a rate of more than 50% per year) (Wikipedia, 2011a).

Some of the important pests on cottonseed that require management for realizing better yields are:

2.1 Locusts and grasshoppers

Identification: there are different species of locusts and grasshoppers that attack oilseed crops especially cotton. Australian plague locust, *Chortoicetes terminifera* (Walker), will not generally damage cotton, although some light damage has been observed on field margins as swarms move through and 'test feed'. The spur-throated locust, *Austracris guttulosa* (Walker), if present in large swarms, can cause damage (Bailey, 2007). The variegated grasshopper, *Zonocerus variegatus* (L.) is the only grasshopper pest in the humid forest zone. It causes damage on most crops, mainly cassava, but also cotton (De Groote, 1997).

Life cycle and damage: Grasshoppers can be occasional early season pests. In late summer and fall, females lay eggs in grassy foothills, on ditch banks, along roadsides and fence rows, in pasture areas. The eggs hatch in spring and the young nymphs feed on nearby plants. Higher temperatures result in faster development of the eggs and nymphal growth. Hot and dry weather conditions favor population increase, while cool and moist conditions decrease population growth. When wild grasses and other plants become dry, the grasshoppers migrate to irrigated croplands. Both nymphs and adults can defoliate crops or chew through the stem weakening the plant which often will fall over at the wound site. They overwinter in egg cases which were deposited in the soil (Collins et al., 2010).

Monitoring: when population densities are high, pest managers can use four 180-degree sweeps with a 15-inch sweep net, which is equivalent to the number of adult (or nymph) grasshoppers per square yard (Knodel et al., 2010).

Economic and action thresholds: action threshold is when 8 or more grasshoppers per square yard occur in the field (Knodel et al., 2010).

Control: Control of annual weeds before grasshopper emergence may reduce grasshopper populations by eliminating alternative food sources for young grasshoppers. *Beauveria bassiana* (Bals.) and *Metarhizium anisopliae* (pathogenic fungi) can be applied for controlling grasshopper population. Scelionid egg parasites and blister beetle are effective biocontrol agents. If the threshold is exceeded apply some insecticides such as acephate at 0.33 - 0.66 lbs, beta-cyfluthrin at 2.4-3.2 fl oz, bifenthrin at 1.6-6.4 fl oz,and esfenvalerate at low rate of 3.9-5.8 fl oz (Knodel et al., 2010).

2.2 *Tetranychus urticae* (Koch) (Acari: Tetranychidae)

Identification: the size of the mite is less than 1 mm and varies in the color. The adults have two typical dark spots on the back so they are named two spotted mite. The female is 0.5 mm long; the male is smaller and slender and they have 4 pairs of legs. The females lay small (less than 0.1 mm in diameter) spherical eggs which are translucent after laying. Larva of reduced size has 3 pairs of legs. The common name, red spider mite, is because they spin silk webs to protect the eggs or colony from predators (Bailey, 2007).

Life cycle and damage: the eggs hatch into the nymph named protonymph, and then turns into deutonymph, afterward adult stage form. *T. urticae* has worldwide distribution and infests to a wide range of plants. The mites convene and feed on the under surface of the leaves causing bronzing, reddening and sometimes desiccation of the leaf. The crop enjoys warm and dry climates which is appropriate for the mites. Yield loss of the mites depends on when mite populations begin to increase and how quickly they increase (Bailey, 2007).

Monitoring: monitor the oldest leaf when plants are very young. As plants grow, choose leaves that are from 3, 4 or 5 nodes below the plant terminal. Begin monitoring at seedling emergence and sample at least weekly. Sample more frequently if the weather conditions are hot and dry. Also, shake the leaves onto a white piece of paper and count the actual number of mites moving around with a hand lens. Eggs and immature stages are difficult to see with the naked eye, so a hand lens should be used (Yvette & Jensen, 2009).

Economic and action thresholds: threshold is when 30% of plants infested through the bulk of the season (squaring to first open boll) (Yvette & Jensen, 2009).

Control: biological control by ladybirds (*Hippodamia convergens* Guerin), big eyed bugs (*Geocoris* spp.), damsel bugs (*Nabis* spp.), lacewings (*Chrysopa* and *Micromus* spp.) and *Phytoseiulus persimilis* (Athias-Henriot) can decrease the population of the mites in early season. Chemical control can be effective by rotation of some acaricides, but only applied when action threshold is observed. Rotation of pesticides is essential because if the same pesticide is used over a prolong period of time they quickly resist to pesticide (Bailey, 2007).

2.3 *Thrips tabaci* (Lind.), *T. palmi* (Karny) and *Scirtothrips dorsalis* (Hood) (Thysanoptera: Thripidae)

Identification: *T. tabaci* adult is about 1 mm long; yellowish-gray to dark-gray and the head has post ocular setae of about the same length as the interocellar setae. The basal segments of the usually seven-segmented antennae are light brown, while the distal segments are dark brown. The prothorax has only two pairs of well developed setae on the posterior margin. The second segment of thoracic has convergent striations on the top surface, and lacks campaniform sensillae. Eggs are microscopic, white or yellow. The nymph ranging from 0.5 to 1.2 mm, white to pale yellow and look like to the adult; just it has short antennas and does not have wings. Pupae are pale yellow to brown, they have short antennae and the wing buds are visible but short and not functional (Rueda, 1995).

T. palmi adults are about 0.8 to 1 mm in length, pale yellow or whitish in color, but with numerous dark setae on the body. A black line, resulting from the juncture of the wings, runs along the back of the body. The slender fringed wings are pale. The hairs or fringe on the anterior edge of the wing are considerably shorter than those on the posterior edge. The egg is colorless to pale white in color and bean-shaped in form. The nymph resembles the adults in general body form though they lack wings and are smaller. The pre pupae and pupae resemble the adults and nymph in form, except that they possess wing pads. The wing pads of the pupae are longer than that of the pre pupae (Capinera, 2010b).

S. dorsalis has small size less than 1 mm in length, yellow coloration, dark antennae with forked sense cones on antennal segments III and IV, antennomeres I-II are pale and III to IX are dark. Also, it has dark striping on the lower abdomen and three distal setae on the lateral margins of abdominal tergites, with pronotal posteromarginal seta II nearly one and a half times the length of I or III, a complete posteromarginal comb on tergite VII; and three ocellar setae with III between posterior ocelli (Wikipedia, 2011d).

Life cycle and damage: *T. tabaci* is known as cotton seedling thrips. Adults prefer to lay their eggs in leaf, cotyledon, or flower tissues. Usually, cotton thrips destroy the cotyledons at first and then other parts of the crop including the bolls. Most damage occurs during early vegetative stage of the crop. Both adult and nymph feed on the under surface of the leaves make them thickened, blistered and bronzed due to continuous feeding. Developing bolls become brown due to development of necrotic patches (Natarajan, 2007; Yvette & Jensen, 2009). *S. dorsalis* can severely damage other oilseed crops such as sunflower.

Monitoring: monitoring should be started from seedling emergence of the crop and number of thrips on 20 - 30 plants counted weekly for every 50 ha (Yvette & Jensen, 2009). Monitor by picking and slapping a leaf on a white index card to see if the tiny specks move around. Open and microscopically examine the plant's growing point for thrips. Plucked growing points can also be dunked in alcohol to dislodge thrips (Bailey, 2007). Also, colored sticky traps can be used for monitoring thrips. Male thrips were most attracted to yellow sticky traps while female thrips were more attracted to pink sticky traps (Yaku et al., 2007).

Economic and action thresholds: The economic threshold from seedling stage until having six true leaves is when 10 adults or nymphs per plant is observed (Yvette & Jensen, 2009).

Control: in moderate infestations, biological control of the thrips by *Anthocoridae, Lygaidae* and predator mites will be effective. Seed treatment with Imidacloprid 70 WS and Chlothianidin (Poncho 600 FS) at 9 ml/kg can protect cottonseeds from early season infestation (Udikeri et al., 2007). However, when high population of thrips recorded systemic insecticides (acetamiprid, imidacloprid and acephate) should be applied (Natarajan, 2007). Aslam et al., (2004) reported that Mospilan, Confidor and Tamaron were highly effective against thrips on cotton.

2.4 *Aphis gossypii* (Glover) (Homoptera: Aphididae)

Identification: *A. gossypii* is commonly referred as cotton aphid. Adults are small (apterate 0.9 - 1.8 mm long and alate 1.1 - 1.8 mm long) and variable in color from greenish brown to orange or dirty yellow with soft body. Nymphs are small sizes of adults.

Life cycle and damage: The females reproduce nymphs parthenogenetically and viviparous which become adults in 7 - 9 days. Cotton aphid is the most common aphid pest infests the cotton seedlings. They are phloem feeders, causing direct leaf crumpling and downward curling with severe attack. Adults and nymphs suck sap from the under surface of the leaves, produce honeydew, indirectly decrease cotton fiber quality and may burn the leaves. Black sooty mould develops on the honeydew interfere photosynthesis of the leaves. Also, they are known as a vector of viruses (Vennila, n.d.). Wilson (2011) reported that aphids transmit plant virus Cotton Bunchy Top appearing in fields across the Australia.

Monitoring: monitor adults and nymphs on the underside of main stem leaves, 3 - 4 nodes below the plant terminal. Presence of ants may indicate presence of aphids. Early detection of aphids is important as they can multiply rapidly. If a high proportion of plants have only the winged form of aphids, recheck within a few days to see if they have settled and young are being produced. Yellow traps are useful for monitoring winged aphids. The presence and abundance of natural enemies should also be recorded (Yvette & Jensen, 2009).

Economic and action thresholds: The economic threshold level in the first open boll stage until harvest is the time if 50% of the plants (counted randomly) are affected or 10% trace amounts of honeydew is present (Yvette & Jensen, 2009).

Control: planting resistant or tolerant varieties which are densely hairy and with stiff leaves are reliable control technique. Seed treatment with *Trichoderma viride* or Imidacloprid 70 WS or Chlothianidin at 9 ml/kg can be effective in reducing population buildup. Aphids are biologically controlled by several species of *coccinellids, chrysopids, syrphids* and the parasitoid *Aphelinus gossypii* (Timberlake). Application of safe and systemic insecticides like Monochrotophos 36SL, Dimethate 30 EC, Endosulphan should be considered if the threshold is exceeded (Natarajan, 2007; Navarajan, 2007; Udikeri et al., 2007).

2.5 *Amrasca devastans* (Distant) or *Amrasca biguttula biguttula* (Ishida) (Homoptera: Cicadellidae)

Identification: adults are usually less than 13 mm long, with slender, tapered bodies of various colors and legs with rows of sharp spines. Eggs are curved and nymphs are flattened, pale yellowish green (Natarajan, 2007).

Life cycle and damage: the common name of the pest is Indian cotton leafhopper or Indian cotton jassid. The female inserts eggs inside leaf veins and after 4 to 11 days the eggs hatch. Nymphs remain confined to the lower surface of leaves during day time. Nymphs' period lasted 7 - 21 days depending on the weather conditions. They have about eleven generations in a year. Both adult and the nymph stages suck the plants sap from under surfaces of the leaves and produce a salivary toxin which impair photosynthesis and cause the edges of the leaves to curl downward. Subsequently, the leaves become yellowish and then redden. Hairy varieties of cotton such as hirsute are less susceptible to the jassids. So, yield loss can be reduced by growing hairy varieties (Natarajan, 2007; Navarajan, 2007). *A. biguttula biguttula* also damage severely other oilseed crops such as sunflower.

Monitoring: monitor the presence of leafhoppers by brushing the foliage; watching for adults and nymphs to jump and fly from plant to plant. Leafhoppers are most sampled with a sweep net. Empty captured jassids into a container with 70% alcohol (or methylated spirits), and express counts as leafhoppers per sweep (one sweep per row) (Bailey, 2007).

Economic and action thresholds: The economic threshold is observing 2-3 adults or nymph per leaf; although for glabrous varieties the threshold may be lowered to one jassid per leaf. Threshold should consider when the margins of the leaves become yellow (Natarajan, 2007).

Control: Natural enemies are not considered to have a significant effect on population of jassids although the parasitoid *Anagrus sp.* has been recorded, but it does not play any significant role in reducing the population. Applying systemic insecticides such as acetamiprid, acephate, so on or seed treatment with imidacloprid or Chlothianidin at 9 ml/kg give good control of jassids (Natarajan, 2007; Navarajan, 2007; Udikeri et al., 2007). also Aslam et al., (2004) declared that Confidor and Mospilan are effective against jassids.

2.6 *Bemisia tabaci* (Genn.) (Hemiptera: Aleyrodidae)

Identification: adults are about 1 mm long with two pairs of white wings and light yellow bodies. Their bodies are covered with waxy powdery materials. Eggs are tiny, oval-shaped, with about 0.25 mm diameter and stand vertically on the leaf surface, eggs at first are white then turn to brown. Nymphs' first instar is 0.3 mm length and the second instar has legs pulled up under its body and other immature stages are sessile. The last nymphal instar develops red eye spots and often is called the pupal stage (Bailey, 2007).

Life cycle and damage: the female lays two types of eggs; one that have not been fertilized and these eggs will result in male offspring. However, fertilized eggs will result in female offspring. Eggs are laid on the under surface of the leaves where the nymphs can suck the plants sap. The incubating time is different due to the season; in the summer eggs hatch after 3-5 days and nymphal period occupy about 9-14 days. However, for winter this period is 5-33 and 17-53 days, respectively. *B. tabaci* has the pupal stage which prolongs approximately 2-8 days and the life cycle ranges from 14-107 days depending on the weather conditions. Whiteflies are found on the under surface of the leaves. They directly feed from the crop, produce honeydew, developing sooty mould and also can transmit viruses (Natarajan, 2007; Navarajan, 2007). *B. tabaci* transmits plant viruses in seven distinct virus groups including: geminiviruses, closteroviruses, carlaviruses, potyviruses, nepoviruses, luteoviruses and a DNA-containing rod-shaped virus. Cotton followed by soybean have been affected by one or more whitefly transmitted viral diseases (Oliveira et al., 2001). They cause contamination of lint through the excretion of honeydew. Their honeydew is considered to be worse than aphid honeydew because it has a lower melting point and during the processing stage, can cause machinery to gum up and overheat (Yvette & Jensen, 2009). Whiteflies can severely damage other oilseed crops such as sunflower.

Monitoring: Monitoring is done by visual counting of nymphs and adults and catching with suction traps or yellow sticky traps. The latter are especially effective in detecting low density populations. Sample the population of larvae per leaf (Ohnesorge & Rapp, 1986).

Economic and action thresholds: the economic threshold is observing 5-10 nymphs or adults per leaf (Navarajan, 2007). Ellsworth & Martinez-Carrillo (2001) declared that based on comparative researches and observations in cotton two adults per leaf in Thailand, 6–8 adults in India, and ca. six adults per leaf in the Sudan were appropriate action thresholds.

Control: *Eretrmocerus mundus* (Mercet) and *Enacarsia* sp. are special parasites of *B. tabaci*. Also, the predators *Amblyseius sp.* (predatary mite), the green lacewing bug (*Chrysoperala sp.*) the coccinellids *Brumus sp., Scymnus sp.* and *Menochilus sp.* play important role in reducing whitefly population. Spray application of systemic insecticides can be effective; also neem oil can control whiteflies build up (Natarajan, 2007; Navarajan, 2007). In addition, Aslam et al., (2004) recommended Mospilan and Actara, as a good control agent against this pest.

2.7 *Oxycarenus hyalinipennis* (Costa) (Hemiptera: Lygaeidae)

Identfication: Adults' measure 4-4.3 mm long; their thorax, head, antennae, and femora are black and wings folded flat on the back and they are translucent white. Males are slightly smaller than females. Nymph has pink to red abdomens. All stages are characterized by a powerful smell when crushed (USDA, 2010).

Life cycle and damage: the cottonseed bug is a major economic threat to cotton. The eggs are laid in open bolls and nymphs can be found in clusters among the lint. This species has five nymphal instars. Adults and nymphs suck sap from mature seeds and leaves of young stems to obtain moisture. These pests prefer inside the bolls and they occasionally go through the leaves of cotton plants or in a single cotton boll, resulting in reduction of seed weight and seed viability in late bolls. A generation from egg to adult can be completed in 20 days, and the pest has three to four generations per year (Smith & Brambila, 2008).

Monitoring: monitor by visual observation and count the number of adults and nymphs. Seed bugs usually found in open bolls and boll opening onwards (Brien, 2010). Also, UV-light traps can be placed in areas related to potential pathways for *O. hyalinipennis* (Derksen et al., 2009).

Economic and action thresholds: No action threshold has been set for *O. hyalinipennis*.

Control: Burning old cotton stalks with bolls may limit future damage by the cottonseed bug. Removal of all weeds or alternative malvaceous host plants near cotton fields is recommended. The cottonseed bug does not normally require control as damage tends to be only to seeds in late bolls. However, Chlorpyrifos-ethyl/hexaflumuron, Dimethoate, Biphenthrin, Endosulfan, Malathion, Methomyl/diflubenzuron and Neem are insecticides available for use against the cottonseed bug (Brien, 2010; USDA, 2010).

2.8 *Creontiades dilutus* (Stal) (Hemiptera: Miridae)

Identfication: Adults of green mirid are about 7mm long, pale green with long antennae. They have sparse setae on their scutellum and pronotum. Females can be distinguished from males by the presence of a median cleft which runs along the mid ventral line of the last abdominal segment. Eggs are about 1.5 mm long, banana shaped, narrowing to a neck below the operculum. Nymphs are the small size of adults; newly hatched nymphs about 1.5-2 mm long and larger nymphs (4th and 5th stage) can be up to 7mm long. They are characterized by their distinctive red tipped antennae (McColl et al., 2011).

Life cycle and damage: Females lay eggs singly within the plant tissue on cotton plants. This species showed a tendency to lay eggs at the distal end of the petioles on the upper one third of the plant. It has four to six nymphal stages. Adults and nymphs cause early season damage to terminals and buds and mid season damage to squares and small bolls; cause blackening, death of young plants terminals and rapid square loss. Bolls that are damaged during the first 10 days of development will be shed, while bolls damaged later than this will be retained but not continue normal development. Black, shiny spots indicate feeding sites on the outside of bolls. When sliced open, warty growths and discolouration of the immature lint can be seen within the boll. So, they cause both direct damage (destruction of terminals, leaves, bolls and branch primordia), and indirect damage (deformed plants) to cotton plants. From first flower until the time when approximately 60% of bolls are 20 days old, the crop is most susceptible to fruit loss that causes yield losses (McColl et al., 2011).

Monitoring: as mirids are very mobile and easily disturbed during monitoring so sample fruit retention and types of plant damage such as tip damage (early season) and boll damage (mid season). Beat sheet or sweep net techniques have proven to be the best sampling methods to find nymphs and adults, respectively (McColl et al., 2011).

Economic and action thresholds: Threshold for *C. dilutus* ranges from 0.5 to 4 mirids per meter of row, depending on the crop stage, sampling technique (visual or beat sheet) and climatic conditions (McColl et al., 2011).

Control: damsel bugs, big-eyed bugs, predatory shield bugs, as well as lynx, night stalker and jumping spiders are known to feed on mirid adults, nymphs and eggs (Yvette & Jensen, 2009). *C. dilutus* is coincidentally controlled in cotton by applications of insecticides targeted at other pests such as *Heliothis* spp. (McColl et al., 2011).

Some other important sucking insect pests of cottonseeds are:
Occasionally green peach aphid, *Myzus persicae* (Sulzer) and cowpea aphid, *Aphis craccivora* (Koch) (Hompotera: Aphididae), infest young cotton. Green vegetable bug, *Nezara viridula* (L.) (Hemiptera: Pentatomidae), pale cotton stainer, *Dysdercus sidae* (Montr.) and red cotton bug, *Dysdercus cingulatus* (F.) (Hemiptera: Pyrrhocoridae) sometimes cause economic damage and heavy crop losses in some regions. Also, brown mirid, *Creontiades pacificus* (Stål) (Hemiptera: Miridae) and cottonseed bug, *Oxycarenus luctuosus* (Montrouzier) (Hemiptera: Lygaeidae) can cause damage to the cotton but the damage of these pests is minor and doesn't require control.

2.9 *Spodoptera littoralis* (Boisduval) (Lepidoptera: Noctuidae)

Identification: The cotton leaf worm adult has a wingspan of 35 to 40 mm. Forewings are brownish with bluish overtones and straw yellow along the median vein. The ocellus is marked by 2 or 3 oblique whitish stripes. The front of the wing tip has a blackish marking, more pronounced in the male. Hind wings are whitish, with a brown front edge. Eggs are spherical and about 0.6 mm long. Neonate larva is pale green with a brownish head; when fully developed (forth instar), it is 35 to 45 mm long. The color of larvae varies from grey to reddish or yellowish, with a median dorsal line bordered on either side by two yellowish-red or grayish stripes, and small yellow dots on each segment. Pupa is about 15 to 20 mm long, brick red color (Vaamonde, 2006).

Life cycle and damage: adults appear in early spring and female lays eggs in clusters and covered with brownish-yellow hairs detached from the abdomen of the female. Most of the clusters are sited on the lower parts of plants. Eggs hatch after 3-4 days and after 2 weeks, pupation takes place in the soil at a depth of 2 to 5 cm. They overwinter as a pupa in the soil. When they are small they feed from the cuticle of the leaves, but if growing they feed from any green part of the plant and also from fruits, resulting in major defoliations. They can also chew the stems, where galleries are drilled. The injuries caused for that pest facilitate the penetration of other pathogens such as fungi, bacteria and etc (Vaamonde, 2006).

Monitoring: detect the presence of adults with monitoring device consisting on insecticide or pheromones traps (3-4 traps per acre). Once the first captures occurs, it is advisable to make a visual estimate by means of a direct sampling on the plants, paying attention to the leaves or other parts of the plant. Count the number of active larvae at the upper and lower surfaces of the leaves and determine the percentage of defoliation (Vaamonde, 2006).

Economic and action thresholds: observation of 35, 45 and 100 egg masses or 0.3, 0.4 and 0.9 larvae per plant is recommended as economic threshold during *S. littoralis* first, second, and third generations, respectively (El-Sherif et al., 1991).

Control: applying azadirachtin and neem oil can control and disrupt growth of insects. Also, sex pheromones can use in the control programs (Martinez & Emden, 2001). Pineda et al., (2007) stated that combination of lethal and sub lethal doses of methoxyfenozide and spinosad might significantly control the population dynamics of *S. littolaris*.

2.10 *Pectinophora gossypiella* (Saunders) (Lepidoptera: Gelechiidae)

Identification: The moth is small and dark brown with blackish bands on the forewings, and hind wings are silvery grey. Eggs are flattened oval and striate about 0.5 mm long, 0.25 mm wide. The larvae are pinkish with dark brown head and about 15 mm length. Pupae are in a thin silken cocoon among the lint inside a seed, in between bracts or in cracks in the soil.

Life cycle and damage: The pink bollworm female lays the eggs on the bolls or in between bracts or on buds and flowers. After 4-25 days, due to the weather conditions, eggs hatch and larvae crawl to the bolls. Larva full grown in 25-35 days and the pupal period is about 6-20 days. The larvae feed early in the season from the inside of the green bolls. Seeds being destroyed in addition to retardation of lint development, lint weakened and stained both inside the boll and in the gin, causing lint yield losses. In addition, infested bolls open prematurely and are invaded by saprophytic fungi.

Monitoring: monitoring can be done with pheromone baited traps as well as by sampling flowers and bolls to record damage levels in cotton fields (Lykouressisa et al., 2005).

Economic and action thresholds: the economic threshold is 8 moths per trap (glossyplure pheromone traps) per day for 3 consecutive days. The number of traps should be 5 per ha or 10% infested flowers or bolls with live larvae (Sabesh, 2004).

Control: Biological control of larvae with *Chelonus* sp., *Camptothlipsis* sp., and *C. blackburni* (Cam.) is effective against overwintering stage. Parasitoids include *Trichogramma brasiliensis* (Ashm.), *Bracon kirkpatricki* (Wilkinson), *Chelonus blackburni* (Cameron); predatory mites such as *Pyemotes ventricosus* (Newport) and *P. herfsi* (Oudemans) are widely reported on reducing bollworm population. Also, entomopathogenic nematodes (Rhabditida: Steinernematidae) such as *Steinernema riobravis* (Cabanillas, Poinar and Raulston) and *Steinernema carpocapsae* (Weiser) may control pests infestations. Applying insecticides is advised if the threshold is exceeded. Application of carbaryl 50 WP, quinalphos 25 EC, profenofos 50 EC, or fumigation of seeds with aluminum phosphide at 18 tablets per 100 cu. m. is recommended (Gouge et al., 1998; Navarajan, 2007; Salama, 1983; Vennila, n.d.).

2.11 *Heliothis armigera* (Hübner) and *Heliothis punctigera* (Wallengren) (Lepidoptera: Noctuidea)

Identification: moth of *H. armigera* is about 35 mm long, with a small pale patch in the dark section of the hind wing; while in *H. punctigera* the dark section is uniform. Eggs of both species are 0.5 mm in diameter, sculptured with longitudinal ribs. *H. armigera* larvae have white hairs around head; medium larvae have saddle of dark pigment on 4th segment from head, and have dark legs. *H. punctigera* larvae have black hairs on the prothorax, dark triangles on the first abdominal segment with light legs. Pupa is smooth surfaced, brown, rounded both anterior and posterior with two tapering parallel spines at posterior tip (Vennila, n.d.).

Life cycle and damage: *H. punctigera* is known as native budworm; and *H. armigera* is the cotton bollworm or corn earworm. *H. armigera* occurs throughout the Africa, Western Europe and Australian pacific region. The adults lay their eggs on young terminal branches of the cotton and after 2 or 3 days the eggs hatch and larvae go through the young leaves and flower buds. So, they burrow fruit and feed from developing seeds and fibers until larval ultimate period. Bollworms have four or five generations and in the last generation burrow into the soil around the base of plant for overwintering and the pupae emerge from the soil in the next spring. *H. punctigera* is similar to *H. armigera*; just there are a few differences between them, such as *H. armigera* usually prefers grass or cereal crops but *H. punctigera* prefers to feed on broadleaf species (e.g. cotton, sunflower, soybean, canola, safflower, linseed and etc.). The diapause is not common in *H. punctigera* (Vennila, n.d.). *H. armigera* also severely damage other oilseeds such as sunflower.

Monitoring: monitore from seedling emergence to maturity of the cotton for the presence of eggs and larvae. Pheromone traps (5 per ha) will detect the presence and intensity of adults (Yvette & Jensen, 2009).

Economic and action thresholds: the economic threshold in early stages of the crop is 2-3 larvae with about 3 mm length/m or 1 larva > 8 mm and for the time when 15-40% of the bolls opened, 5 larvae with 3 mm length/m or 2 larva > 8 mm/m (Yvette & Jensen, 2009).

Control: mechanical cultivation of the soil at the end of the season destroys the tunnels and shelters made by the larvae and can kill over 90% of the pupae in the soil. *Trichogramma* spp., *Compoletis chloridae* (Uchida) can release against early instar larvae and Tachinids: *Carcelia illota* (Curran), *Goniopthalmus halli* (Mesnil.) and *Paleorixa laxa* (Curran) have been recorded on late larval instars of *H. armigera* (Vennila, n.d.). Spraying insecticides can be applied by consulting the field guide because *H. armigera* indicates resistance to insecticides (Salama, 1983). Recently, planting of transgenic *Bacillus thuringiensis* (Bt) cotton seeds is approved (Menon & Jayaraman, 2002).

2.12 *Earias insulana* (Boisd.) and *Earias vittella* (F.) (Lepidoptera: Noctuidae)

Identification: *E. vittella* moth has green forewings with a white streak on each of them whereas that of *E. insulana* is completely green. The eggs are less than 0.5 mm diameter, crown-shaped, sculptured and blue. The larva of *E. vittella* is brownish with a longitudinal white stripe on the dorsal side and without finger-shaped processes on its body. Larva is cream with orange dots on the prothroax. The boat shaped tough silken cocoon is dirty white brownish (Navarajan, 2007).

Life cycle and damage: They are known as spotted bollworm. Females lay eggs singly on most parts of the cotton such as flower buds, bolls, peduncles, bracts, leaf axils and veins on the under surface of leaf and after 3 days eggs hatch. The larval period is about 10-12 days. In the early vegetative stage of the crop the larvae cause damage to the terminal bud of the shoots and channel downwards or into inter node which redound to drying the shoot. Afterward, in the later stages of the crop growth; the larvae migrate to the buds, flowers, bolls and fresh parts of the plants and damaged them. Tunnel in bolls is often from below, angled to the peduncle. Larvae do not confine their feeding to a complete single boll; hence damage is disproportionate to their numbers. So, heavy shedding of early formed flower buds is observed in cotton fields and the lint from attacked bolls will not be clean. The period of pupae is 7-10 days. The total life cycle ranges from 20 to 22 days (Navarajan, 2007).

Monitoring: monitor the population fluctuation by installation synthetic sex pheromone traps. Also, determine larval infestation in bolls of cotton (Qureshi & Ahmed, 1991)

Economic and action thresholds: the economic threshold is differ from region to region, Navarajan (2007) stated 5% damaged fruiting bodies or one larva per plant from 20

Table — Impact of insecticides and miticides on predators, parasitoids and bees in cotton

By Lewis Wilson[32], Simone Heimoana[32], Robert Mensah[32], Moazzem Khan[32], Martin Dillon[32], Mark Wade[4], Brad Scholz[4], DavidMurray[32], Richard Lloyd[3], Richard Sequeira[3], Paul DeBarro[6] Viliami Heimoana[6] and Jonathan Holloway[6]. [32] CSIRO Plant Industry. [2] I&I NSW. [3] DEEDI. [4] CSIRO Entomology. [6] Formerly CSIRO Entomology. [6] Formerly I&I NSW. [7] Cotton CRC.

Insecticides (in increasing rank order of impact on beneficials)	Rate (g a.i./ha)	Target pest(s)						Persistence[8]	Overall Ranking[9]	Total[1] Predatory beetles	Red and blue beetle	Minute 2-spotted lady beetle	Other lady beetles	Total[2] Predatory bugs	Damsel bugs	Big-eyed bugs	Other predatory bugs	Apple dimpling bug	Lacewing adults	Spiders	Total wasps	Eretmocerus 19	Trichogramma	Ants	Thrips	Mite (resurgence)	Aphid (resurgence)	Helicoverpa (resurgence)	Toxicity to bees[14]
		Helicoverpa	Mites	Mirids	Aphids	Thrips	Silverleaf whitefly																						
Bt[11]		✓						very short	very low	VL	VL	VL	VL	VL	VL	VL	VL	VL	VL	VL	VL	VL	VL	VL	VL	–	–	–	VL
NP Virus		✓						very short	very low	VL	VL	VL	VL	VL	VL	VL	VL	VL	VL	VL	VL	VL	VL	VL	VL	–	–	–	VL
Pirimicarb	250				✓			short	very low	VL	VL	VL	VL	VL	VL	M	VL	VL	VL	VL	VL	M	M	VL	VL	–	–	–	VL
PSO (Canopy)[16]	2%	✓			✓			short	very low	VL	VL	VL	VL	VL	VL	VL	VL	VL	VL	VL	VL	VL	VL	VL	VL	–	–	–	VL
Methoxyfenozide	400	✓						medium–long	very low	VL	VL	VL	VL	VL	VL	VL	VL	VL	VL	VL	VL	VL	VL	VL	VL	–	–	–	VL
Pyriproxyfen	50						✓	long	very low	M	–	M	M	VL	VL	–	VL	–	VL	M	VL	–	VL	VL	VL	–	–	–	–
Boxazole	38.5		✓					short	low	VL	VL	H	VL	VL	VL	VL	L	H	L	VL	L	L	L	L	L	–	–	–	VL
Indoxacarb (low)	60			✓				medium	low	L	M	H	M	M	M	L	L	L	M	L	L	L	L	H	L	–	+ve	–	–
Indoxacarb (low + salt)	60			✓	✓			medium	low	L	M	H	M	M	M	L	L	L	M	L	L	L	L	H	L	–	+ve	–	–
Indoxacarb (low + Canopy)	60			✓				medium	low	L	–	H	M	M	M	L	L	L	M	L	L	L	L	H	L	–	+ve	–	–
Rynaxypyr	52.5	✓						long	low	L	M	–	VL	L	L	–	L	VL	VH	VL	L	M	VL	–	L	–	+ve	+ve	VL
Dicofol[3]	960		✓					long	low	L	–	H	M	M	M	–	L	M	M	L	L	L	L	–	L	–	–	–	–
Amorphous silica[17]	2500	✓			✓		✓	short	low	L	M	–	VL	M	M	VL	L	–	L	VL	M	M	L	M	L	+ve	–	–	–
Spinosad	96						✓	medium	low	M	H	H	M	M	M	L	H	M	M	L	H	H	H	H	H	–	–	–	H[15]
Diafenthiuron	350		✓		✓			medium	low	H	H	H	H	M	M	L	H	–	L	M	H	L	VL	H	VL	–	–	+ve	M
Pymetrozine	150				✓			short	low	L	L	VH	VL	L	L	VL	VL	VL	VL	L	L	L	VL	VL	VL	–	–	–	VL
Fipronil (v. low)	8			✓				medium	low	L	–	M	L	L	L	–	L	VH	L	L	H	–	–	VH	L	+ve	+ve	–	VH
Fipronil (v. low + salt)	8			✓				medium	low	L	–	M	L	L	L	–	L	VH	L	L	H	–	–	VH	L	+ve	+ve	–	VH
Indoxacarb[13]	127.5	✓		✓				medium	low	H[14]	VH	VH	VL	M	M	M	VL	H	H	M	H	H	VL	VH	M	+ve	+ve	–	H[15]
Abamectin	5.4	✓[6]	✓					medium	moderate	M	M	M	M	M	M	M	M	H	M	M	M	M	M	H	M	+ve	+ve	–	H
Emamectin	8.4	✓						medium	moderate	M	VL	M	VL	M	M	H	H	H	VL	M	M	H	M	VL	M	–	–	–	H

Product	Rate	Persistence[8]	Impact											Bees[14]	Resurgence[12]
Dimethoate (low)	80	short	moderate	M	H		H	M	M	M	M	M	+ve	+ve	H
Dimethoate (low + salt)	80	short	moderate	M	H		H	M	M	M	M	M	+ve	+ve	H[15]
Endosulfan (low)	367.5	medium	moderate	VL	H	M	H	H	VL	L	M	VH	–	+ve	M[15]
Propargite	1500	medium	moderate	M	M		M	M		M	M	M	+ve	+ve	VH
Acetamiprid	22.5	medium	moderate	M	VH		VH	VH		H	M	H	–	–	M[15]
Clothianidin (low)	25	medium	moderate	VL	VH		VH	VL		M	H	VH	–	+ve	H
Amitraz	400	medium	moderate	H	M		M	M		M	M	M	–	–	VH
Fipronil (low)	12.5	medium	moderate	L	VH	L	H	VH		H	VH	VH	+ve	+ve	VH
Chlorfenapyr (low)	200	medium	moderate	M	M	M	H	M		M	M	M	+ve	–	M
Thiamethoxam	100	medium	moderate	H	H	H	H	M		L	M	H	+ve	+ve	VH
Endosulfan (high)	735	medium	moderate	M	VH	M	M	H		H	VH	VH	–	–	M[15]
Fipronil (high)	25	medium	moderate	VL	VH	M	M	H		H	VH	VH	+ve	+ve	VH
Imidacloprid	49	medium	moderate	L	VH		H	H		M	VH	VH	+ve	+ve	M
Clothianidin (high)	50	medium	moderate	VL	VH	VH	H	H		H	VH	VH	+ve	+ve	H
Methomyl	169	very short	high	VH	VH		H	H		H	H	H	–	–	H[15]
Thiodicarb	750	long	high	M	VH	VH	H	H	M	H	H	H	+ve	+ve	M[15]
Dimethoate (high)	200	short	high	M	H		M	H		M	M	M	+ve	+ve	H
Chlorfenapyr (high)	400	medium	high	M	H		M	H		H	M	M	+ve	–	H
OPs[5]		short–medium	high							H	H	H	–	–	H
Carbaryl[11]		short	high										–	–	
Pyrethroids[4]		long	very high	VH	H	VH	VH	VH		VH	H	VH	+ve	+ve	H

Fig. 2. Some effective pesticides against cotton pests and their efficiency on parasitoids, predators and bees (Yvette & Jensen, 2009).

1. Total predatory beetles – ladybeetles, red and blue beetles, other predatory beetles.
2. Total predatory bugs – big-eyed bugs, minute pirate bugs, brown smudge bugs, glossy shield bug, predatory shield bug, damsel bug, assassin bug, apple-dimpling bug
3. Information; Citrus pests and their natural enemies, edited by Dan Smith, University of California Statewide IPM project, Cotton, Selectivity and persistence of key cotton insecticides and miticides.
4. Pyrethroids; alpha-cypermethrin, cypermethrin, beta-cyfluthrin, cyfluthrin, bifenthrin, fenvalerate, esfenvalerate, deltamethrin, lambda cyhalothrin
5. Organophosphates; omethoate, monocrotophos, profenofos, chlorpyrifos, chlorpyrifos-methyl, azinophos ethyl, methidathion, parathion-methyl, thiometon
6. Helicoverpa punctigera only.
7. Bifenthrin is registered for mite and silverleaf whitefly control; alpha-cypermethrin, beta-cyfluthrin, bifenthrin, deltamethrin and lambda-cyhalothrin are registered for control of mirids.
8. Persistence of pest control; short, less than 3 days; medium, 3-7 days; long, greater than 10 days.
9. Suppression of mites and aphids only.
10. Impact rating (% reduction in beneficials following application, based on scores for the major beneficial groups) VL (very low), less than 10%; L (low), 10-20%; M (moderate), 20-40%; H (high), 40-60%; VH (very high), > 60%. A '-' indicates no data available for specific local species.
11. Bacillus thuringiensis.
12. Pest resurgence is +ve if repeated applications of a particular product are likely to increase the risk of pest outbreaks or resurgence. Similarly sequential applications of products with a high pest resurgence rating will increase the risk of outbreaks or resurgence of the particular pest species.
13. Very high impact on minute two-spotted ladybeetle and other ladybeetles for wet spray, moderate impact for dried spray.
14. Data Source: British Crop Protection Council 2003. The Pesticide Manual: A World Compendium (Thirteenth Edition) Where LD50 data is not available impacts are based on comments and descriptions. Where LD50 data is available impacts are based on the following scale: very low = LD50 (48h) > 100 ug/bee, low = LD50 (48h) > 100 ug/bee, moderate = LD50 (48h) < 10 ug/bee, high = LD50 (48h) < 1 ug/bee, very high = LD50 (48h) < 0.1 ug/bee. Refer to the Protecting Bees section in this booklet.
15. Wet residue of these products is toxic to bees, however, applying the products in the early evening when bees are not foraging will allow spray to dry, reducing risk to bees the following day.
16. May reduce survival of ladybeetle larvae - rating of M for this group.
17. May be detrimental to eggs and early stages of many insects, generally low toxicity to adults; and later stages.
18. Will not control organophosphate-resistant pests (e.g. mites, some cotton aphid (Aphis gossypii) populations
19. Rankings for Eretmocerus based on data for E. mundus (P. De Barro, CSIRO, unpublished) and for E. eremicus (Koppert B.V., The Netherlands (http://side-effects.koppert.nl/#))

DISCLAIMER: Information provided is based on the current best information available from research data. Users of these products should check the label for further details of rate, pest spectrum, safe handling and application. Further information on the products can be obtained from the manufacturer.

randomly counted plants. However in the other study, Qureshi & Ahmed (1991) declared that a mean trap catches of 9-12 moths per night was associated with economic injury level and the economic threshold is 10% larval damage.

Control: application of insecticides such as phosalone 35 EC, carbaryl 50 WP, endosulfan 35 EC, monocrotophos 40 SC, thiodicarb 75 WP, or profenofos 50 EC are advised. Also, the synthetic pyrethroids, fenvalerate, deltamethrin and permethrin has been reported to be effective against bollworms (Navarajan, 2007).

3. Groundnuts

Groundnuts contain seeds of:

Arachis villosulicarpa Hoehne: known as groundnut; cultivated in some states of Brazil.
Vigna subterranea L.: known as Bambara groundnut; originated in West Africa and a traditional food plant in Africa.
Macrotyloma geocarpum Harms: known as geocarpa groundnut, Hausa groundnut, or Kersting's groundnut, in Sub-Saharan Africa.
Arachis hypogaea L.: known as peanut, earthnuts, ground nuts, goober peas, monkey nuts, pygmy nuts and pig nuts. About 41.5% of world production is in China, 18.2% in India and 6.8% in the United States of America.
Also, roots and: *Apios americana* Medik.: known as potato bean, hopniss, Indian potato or groundnut, native to eastern North America.
Conopodium majus Gouan: known as kippernut, cipernut, arnut, jarnut, hawknut, earth chestnut, groundnut, and earthnut, native to Europe and parts of North Africa.
Panax spp.: is in parts of north America and eastern Asia (Wikipedia, 2010).
A few of the insect pests that cause economic losses to groundnut are introduced here:

3.1 *Tetranychus cinnabarinus* (Bois.) (Acari: Tetranychidae)

Identification: females are reddish, and more or less elliptical. The males are slightly smaller and wedge shaped. They have a black spot on either side of their relatively colorless bodies. Eggs are spherical, shiny, straw colored. Larvae are slightly larger than the egg, pinkish, and have three pairs of legs. They have two nymphal stages, the protonymph and deutonymph. The nymphal stage differs from the larval stage by being slightly larger, reddish or greenish, and having 4 pairs of legs (Mau & Kessing, 2007a).

Life cycle and damage: the females lay eggs singly on the underside of the leaf or attached to the silken webs spun which hatch in 3 days. The larval stage lasts a short time, perhaps a day and afterward formed the nymphs. Nymphal stage lasts about 4 days. Spider mites feed on the underside of the uppermost leaves. They can be very serious pests during extended dry periods (Mau & Kessing, 2007a). Resistance of peanut, and wild species of *Arachis* has been declared to the two spotted spider mite in the USA (Lynch, 1990).

Monitoring: monitor for areas of light colored (chlorotic) plants, especially along field edges. Look under the top leaves to see if mites are present (Linker et al., n.d.).

Economic and action thresholds: No threshold has been set for mites (Linker et al., n.d.).

Control: mites quickly resist to pesticide so consult with county Cooperative Service Extension agent which pesticides are associated with mite problems.

3.2 *Balclutha hortensis* (Lindberg) and *Empoasca kerri* (Pruthi) (Homoptera Cicadellidae)

Identification: *B. hortensis* adults are small and have a well developed appendix on the tegmen. Most species of *Balclutha* genus are green coloured although some of them are pink or red (NSW, 2001). *E. kerri* adults are yellowish green in color, wedge shaped and walked diagonally in a characteristic manner (Khanpara, 2011).

Life cycle and damage: the females insert eggs into the leaf tissue close to the midrib or into the petiole. The eggs hatch in a week and in 10 days nymphs change to the adult form. The nymphal stage passed through five instars. Both nymphs and adults suck sap from central surface of leaves; inject toxins resulting in whitening of veins and chlorotic patches especially at the tips of leaflets. This type of feeding cause 'V' shaped yellowing. Heavily attacked crops look yellow and give a scorched appearance known as 'hopper burn'. Also, stunting of crop growth is observed where they are endemic. life cycle is about 18-30 days in the case of male, while it is 20-34 days for female at an average temperature of 31.30±1.96°C and 73.96±3.02% r.h. (Ghewande & Nandagop, 1997; Khanpara, 2011). In the USA plant resistance has been confirmed to the leafhoppers (Lynch, 1990; Nandagopal & Reddy, 1987).

Monitoring: monitoring should be done weekly from 3 weeks after emergence ($R1$) up to maturity ($R9$) when the plants had no leafhoppers left on them. D-vac suction sampler can use for determining the number of adults and nymphs (select 10 sites in the field and five plants randomly per sample site). Also, determine the percentage of leaves affected by hopper burn (Linker et al., n.d.).

Economic and action thresholds: the economic threshold level is when 25% of the leaves are damaged or the presence of 5-10 nymphs or adults per plant. If the field is to be sprayed with fungicide, 15% threshold can be used (Linker et al., n.d.).

Control: planting tolerant varieties; crop rotation with non host crop; intercropping with pearl millet; avoidance of groundnut-castor inter crop; and once irrigation to avoid prolonged mid season drought are cultural techniques for preventing pests infestations. Applying safe insecticides such as Dimethoate 30 EC, or Monochrotophos 36SL can be effective against jassids (TIFP, n.d.-a).

3.3 *Caliothrips indicus* (Bagn.); *Thrips palmi* (Karny); *Scrirtothrips dorsalis* (Hood); *Frankliniella schultzei* (Trybom) (Thysanoptera: Thripidae)

Identification: in the case of *C. indicus* both adults and nymphs are dark colored with fringed wings. For *F. schultzei*, dults are dark colored with fringed wings but nymphs are yellowish. For identification *T. palmi* and *S. dorsalis* refer to the cotton (Krishworld, n.d.).

Life cycle and damage: symptoms of damage is different as tender leaves show yellowish green patches on the upper surface; brown necrotic areas and silvery sheen on the lower surface. Severe infestations cause stunted plants. For *C. indicus*, lower leaves showing white spots or streaks intermingled with black excreta on the surface. In *F. schultzei* damage, young and terminal leaves showing white scars. They also transmit peanut bud necrosis. Peanut and its wild species show resistance to thrips species in the USA (TNAU, n.d.-c).

Monitoring: Colored sticky traps or water traps are useful for monitoring thrips. The color spectrum of sticky traps influenced their efficacy. Blue, yellow or white colors are used especially and it seems bright colors attract more thrips than darker ones.

Economic and action thresholds: economic threshold level is when 25% of the leaves are damaged and live thrips are found in the field or 5 thrips per terminal shoot (Bhubaneswar, 2008; Linker et al., n.d.).

Control: Spraying Monochrotophos 36 WSC, Dimethoate 30 EC, Methyldemeton 25 EC, or Monocrotophos 320 ml mixed with neem oil, 1 lit and 1 kg soap powder mixed in 200 lit of water twice at 10 days interval can be effective against thrips infestations (TNAU, n.d.-c).

3.4 *Aphis craccivora* (Koch) (Homoptera: Aphididae)

Identification: *A. craccivora* is known as groundnut aphid or cowpea aphid. The adult is about 2 mm long, pear shaped, green greenish brown or greenish black in color. The nymphs are dark brown and turns to shiny dark. Adults are mostly wingless but few winged forms also seen (TIFP, n.d.-a).

Life cycle and damage: Both nymphs and adults feed on the growing tips, tender foliage, flowers and growing pegs causing stunting and distortion of the foliage and stems. They excrete honeydew on which sooty molds develop. Aphids are also known to transmit peanut stripe virus and groundnut rosette virus complex. They have 12-14 generations per year (TIFP, n.d.-a). In the USA Peanut and its wild species show resistance to the groundnut aphid (Lynch, 1990). Bottenberg & Subrahmanyam (1997) stated that aphids in central Malawi were first recorded when the crop was at the flowering stage, but after 2 weeks the population of aphids decreased and the infestation ranged from 6 to 32%.

Monitoring: monitoring should be done during seedling, flowering and pegging stages of the crop. Monitor undersides of the leaves and yellow traps can use for monitoring aphids.

Economic and action thresholds: on average recorded in 1986, *A. craccivora* was of economic importance with economic threshold of 22.3 aphids per plant. However, the economic damage of aphids varies with the stage of plant, with most damage caused if aphids infest the growing points of groundnut early in the plant's development (CABI, 2011).

Control: releasing *Cheilomenes sexmaculata* (F.) or other biological control agents such as flower bugs (Anthocorids), and etc. would be effective in controlling aphids (Jasani, 2009).

3.5 *Anisolabis stalli* (Dohrn) (Dermaptera: Forficulidae)

Identification: Pod borer adults are dark brown to black with forceps like caudal cerci and white leg joints. Nymphs are white in early stages and later turns brown (TNAU, n.d.-c).

Life cycle and damage: young pods showing bore holes plugged with excreta. Sand particles or discolored pulps is also observed and the pods lose their kernel (TNAU, n.d.-c).

Monitoring: monitoring can be accomplished with wheat bran or oat meal baits or traps. Likewise, traps take advantage of the natural tendency of earwigs to hide in crevices and can be used to detect presence of earwigs, and to estimate abundance (Capinera, 2009).

Economic and action thresholds: No action threshold has been set for pod borers.

Control: Application of Malathion 5D 25 kg/ha or Endosulfan 4D 25 kg/ha to the soil prior to sowing in endemic areas. Forty days after sowing, the application of insecticides should be repeated (TNAU, n.d.-c).

3.6 *Odontotermes obesus* (Rambur) (Isoptera: Termitidae)

Identification: termites are social insects, live in territorial, in distinct castes, workers, kings and queen. Workers are smaller; have a soft, white body and a brown head (Jasani, 2009).

Life cycle and damage: termites are one of the major soil pests of groundnut. However, in the USA peanut and its wild species show resistance to the genus *Odontotermes* (Lynch, 1990). Termites favor red and sandy soils and lay eggs on plants or in the soil. They penetrate, hollow out the tap root and feed on the roots of the groundnut. The attack continues with boring holes into pods and damages the seed. It removes the soft corky tissue from between the veins of pods causing scarification, weaken the shells, and make them liable to entry and growth of *Aspergillus flavus* that produces aflotoxins (Jasani, 2009; Umeh et al., 2001).

Monitoring: for determining the presence of termites in the field insert two pieces of wood into a hole in the ground as monitoring devices at fixed intervals (typically 10 to 20 feet apart). If termites are found in the monitoring station, woods are replaced with a perforated plastic tube containing baits. After termites are no longer found in installed bait tubes, the baits are once again replaced with untreated wood pieces and monitoring continues.

Economic and action thresholds: No action threshold has been set for termites.

Control: cultural control techniques for termites are: digging the territorial and destruction of the queen and harvesting the groundnuts as soon as they are matured. Termites can be controlled mechanically by avoiding physical loss of the crop during harvesting, destruction of debris, nests and queen. Applying insecticides like Chlorpyriphos 20 EC, Lindane 1.3% or Chlorpyriphos dust in soil before sowing may reduce termite damage (Jasani, 2009).

3.7 *Holotrichia consanguinea* (Blanchard), *Holotrichia serrata* (F.) and *Sphenoptera indica* (Laporte & Gory) (Coleoptera: Scarabaeidae)

Identification: The adult of *S. indica* is shiny beetle, about 10 mm long and 3 mm wide. The larvae are C shaped, slow movers having globular head and elongated, dorso-ventrally flattened body. Population takes place in the larval tunnel. Adults of *Holotrichia* spp. are about 18-20 mm long and 7-9 mm wide. The eggs are white, almost round. The young grubs are translucent, white and 5 mm long (Jasani, 2009).

Life cycle and damage: grubs are one of the major soil pests of groundnut. They are polyphagus and both adults and larvae are damaging stages. The females lay eggs singly on the main stem. The grubs live in soil and remain active, feed on the functional roots of the plant, leaving behind only tap root. They also burrow into the stem, close to the soil surface and kill the plant. Grub infested plants turn pale, leaves and branches drop down, the plant withers and can be easily uprooted. It ultimately dies off resulting in patchy crop growth (Jasani, 2009; Umeh et al., 2001).

Monitoring: install light traps in the field; with the onset of rains the beetles come out of the soil and attract to the light so count the number of beetles per day. Also, dig 10 pits per ha in $100 \times 100 \times 20$ cm and count the number of beetles per pit every day (Jasani, 2009).

Economic and action thresholds: the economic threshold in south-central India for white grubs is observation of 0.14 larvae m⁻² (Anitha et al., 2005).

Control: planting tolerant varieties; deep ploughing in the summer; manual destruction of infested plant stems; treatment of the seeds before sowing with Chloropyriphos can control grubs population and their damage. The seeds can further be treated with fungicides like Thiram. Applying carbofuran granules in the planting row can be effective (Jasani, 2009).

3.8 *Diabrotica undecimpunctata howardi* (Barber) (Coleoptera: Chrysomelidae)

Identification: adults are greenish-yellow beetles and have 12 irregular black spots on their backs. Eggs are about 0.5 mm long, oval and yellow. The damage to peanuts, however, is caused by the whitish-clear tender looking larvae. The larvae are 12-19 mm long and have a dark head, and a small dark spot at the rear (Mossler & Aerts, 2009).

Life cycle and damage: The southern corn rootworm female lays 200-1200 eggs in the soil near host plants. They usually hatch in 1-2 weeks, depending on the temperature. Larvae damage peanuts by feeding on the pods and pegs in the soil. These pests tend to be a problem more often in damp weather and on heavier type soils; however, it can occur in sandy soils. Larvae develop over about 2-4 weeks. They then pupate in the soil and adults emerge about 1-2 weeks later. Adults may also be vectors of plant pathogenic bacteria and viruses, e.g. *Pantoea stewartii* subsp. *stewartii* (synonym: *Erwinia stewartii*) (Linker et al., n.d.).

Monitoring: monitor after pegging with a hand trowel or small spade to unearth the nuts under 2-3 row feet of plants (without digging up the plants). Monitor randomly, but check low, moist spots or areas of heavier soil where plants are most likely to show signs of damage. Examine the pegs and pods for small holes and signs of boring (Linker et al., n.d.).

Economic and action thresholds: no threshold has been set for rootworm. However, if 10% of samples showed live larvae or fresh damage, treatment is advised (Linker et al., n.d.).

Control: management of this pest can be quite frustrating because it feeds underground. Early-planted fields or crop rotated ones are at less risk from damage. Apply insecticides such as Chlorpyrifos if the threshold is exceeded (Mossler & Aerts, 2009).

3.9 *Stomopteryx submissella* (Frey) and *Aproaerema modicella* (Deventer) (Lepidoptera: Gelechiidae)

Identification: The moth of *S. submissella* is brownish gray, 6 mm long with 10 mm wing span. The eggs are shiny white. The larvae are about 1 mm long at first and enlarge as the larvae grow. Pupation takes place in the webbing (Jasani, 2009). *A. modicella* moths are brownish grey, 6 mm long with 10 mm wing span. Forewings have white spot on the costal margin. The larvae are green in color with dark head and prothroax (TNAU, n.d.-c).

Life cycle and damage: leaf miner is one of the most important pests of groundnut. The females lay eggs singly underside of the leaflets. Young larvae initially mine into the leaflets, feed on the mesophyll and form small brown blotches on the leaf. Afterward larvae web the leaflets together and feed on them, remaining within the folds (Jasani, 2009). In the USA plant resistance has been confirmed to the *A. modicella* (Lynch, 1990).

Monitoring: monitor moths by pheromone trap, 1/ha or light trap 12/ha. Also, count the number of larvae on randomly selected plants in the field (Jasani, 2009).

Economic and action thresholds: Economic threshold is 2-3 larvae per plant or 10% leaflets damaged in central whorl (Bhubaneswar, 2008; Jasani, 2009).

Control: planting of cowpea or soybean as trap crop; crop rotation with non leguminous crop is advised and rotation with leguminous crops should be avoided. There are some resistant or tolerant varieties that can be planted. Egg masses and early instars larvae can be collected and destroyed manually. Also, installation of pheromone traps for mass trapping is one of the mechanical control techniques of leaf minor. *Trichogramma Chilonis* (Ishii) is special biological control agent of the minor in groundnut. *Pennisetum glacum* (L.) enhanced the parasitoid *Goniozus* spp. on leaf miner. Applying insecticides is recommended if the insect population crosses the ETL. Carbaryl 50WP, Quinalphos 25 EC, Methyldemeton 25 EC or Dimethoate 30 EC are effective against leaf minor (Jasani, 2009; TNAU, n.d.-c).

3.10 *Spilosoma (Diacrisia) obliqua* (Walker) (Lepidoptera: Arctiidae)

Identification: Bihar hairy caterpillar moth is brown with a 40-50 mm wing span and a red abdomen. Eggs are light green spherical in clusters on the underside of leaves. The larvae are covered with long yellowish to black hairs and are up to 5 cm long. The pupa forms a thin silken cocoon by interwoven shed hairs of the larvae (Jasani, 2009; TNAU, n.d.-c).

Life cycle and damage: each female lays 533-1287 eggs and the eggs hatch after about 6-9 days. Larval and pupal stages lasted 34-45 and 16-22 days, respectively, giving a total life cycle of 59-76 days. Young larvae feed gregariously on the under surface of the leaves and cause loss by way of defoliation. Sometimes, after defoliated the crop larvae feed on the capsules. In severe cases only stems are left behind. Pupation takes place in the soil under dry foliage and debris where the pupae overwintering (Jasani, 2009; TNAU, n.d.-c). They also severely damage other oilseeds such as sunflower.

Monitoring: monitor the flight intensity of the male moth by using pheromone traps 5/ha and determine the percentage of crop defoliation (TNAU, n.d.-c).

Economic and action thresholds: economic threshold is when 20-25% defoliation is observed (Bhubaneswar, 2008).

Control: deep ploughing in the pre-monsoon (two/three times) will expose the hibernating pupae to sunlight and predatory birds. Removal and destruction of alternate wild hosts which is the harbor of caterpillars can be effective method of controlling *S. obliqua*. Also, planting trap crops like cowpea, castor and Jatropha on field bunds to attract the caterpillars may be efficient. Setting up bonfires on field bunds during night, mass collection and destruction of eggs and emerged larvae are mechanical ways of controlling hairy caterpillars. Spraying B.t is also advocated at 1 kg/ha where mulberry is not grown. Application of insecticides should be considered if the threshold is exceeded. Dust Lindan 1.3% and Fanvalerate 0.4% can use in the early stage of larvae. In addition, spraying Quinalphos 25 EC, Chlorpyriphos 20 EC or Endosulfan 35 EC recommended when the caterpillars are younger (Jasani, 2009; TNAU, n.d.-c).

3.11 *Amsacta albistriga* (Walker) (Lepidoptera: Arctiidae)

Identification: the moths emerge from the soil at the onset of the south-west monsoon. They are brownish-white moths with a 40-50 mm wing span. The larvae are light brown and turn reddish as they grow. They are haired and are up to 5 cm long (Jasani, 2009).

Life cycle and damage: Females lay around 800-1000 eggs in clusters of 50-100 on the host plants. The eggs hatch in 2-3 days and tiny first instar caterpillars remain under the cover of natural vegetation for about 8-10 days. A week to 10 days old caterpillars spreads to the fields and start feeding. Red hairy caterpillars cause defoliation of the crop as they are voracious feeders and often migrate from one field to another devastating whatever crops come their way. After about 30-40 days of feeding the larvae burrow into soil, usually in the undisturbed soil of field or non-cropped areas and pupate for diapauses. They have one generation in a year (Jasani, 2009; TNAU, n.d.-c).

Monitoring: monitor by erection of 12 light traps per ha for 20-45 days (Jasani, 2009).

Economic and action thresholds: Economic threshold is when 15-20% of plants are affected.

Control: there are different methods for managing this pest. The cultural techniques are: deep ploughing in summer to expose the pupae to predatory birds; early sowing to escape crop from pest infestations; mechanical weeding at 15-20 days after sowing; and intercropping one row of castor for every 5 or 6 rows of groundnut. Crop rotation with sorghum, pearl millet or maize should be followed. Vegetative traps utilizing jatropa (wild castor) or ipomoea prevent the migration of the grown up larvae. Irrigate once to avoid prolonged mid season drought to prevent pre-harvest infestation. Install light traps in endemic areas, collect then kill the moths. Collect and destruct egg masses in the fields around light trap areas. Red hairy caterpillar can be controlled biologically by releasing *Coccinella* sp., and parasitoids like *Bracon hebetor* (Say), *Chelonus* spp. Also, spraying A-NPV and B.t can reduce the *A. albistriga* population and if the threshold is exceeded applying insecticides is recommended. Methyl parathion 2%, Fanvalerate 0.4%, Endosulfan 35 EC, Quinalphos 25 EC, Nuvan (76%) can control full grown insect pests (TNAU, n.d.-c).

3.12 *Spodoptera litura* (F.) (Lepidoptera: Noctuidae)

Identification: Tobacco bud worm moths are light brown with a wing span of about 30 mm and mottled forewings. The eggs are laid in the masses about 4 × 7 mm and appear golden brown on the upper surface of leaves. Young larvae are light green in color. Full grown larvae are stout, cylindrical and pale greenish brown with dark markings. The pupae are reddish brown and are found in the soil close to the plant (Jasani, 2009).

Life cycle and damage: females lay around 2000 eggs on the abaxial surface of groundnut leaves, in batches of 200-300 each. The larval and pupal periods take about 20 and 7-10 days, respectively. Freshly hatched larvae feed gregariously, scraping the chlorophyll, and they disperse very soon. Larvae feed mostly during night time. Later stages feed voraciously on the foliage at night, hiding usually in the soil around the base of the plants during the day. Sometimes the feeding is so heavy that only petioles and branches are left behind. In light soil, caterpillar bores into the pods. The total life cycle is completed in 30 days and there may be as many as 12 generations annually in southern India (Jasani, 2009; Mallikarjuna et al., 2004). Peanut and its wild species have three flavonoids chlorogenic acid, quercetin and rutin which are involved in the components of resistance to *Spodoptera* species (Mallikarjuna et al., 2004). They also severely damage other oilseeds such as sunflower.

Monitoring: pheromone trap can use for monitoring *S. littoralis* (5 per ha).

Economic and action thresholds: economic threshold for this pest is 20-25% defoliation of the crop (Bhubaneswar, 2008).

Control: some cultural techniques for controlling *S. litura* are: deep summer ploughing; early sowing to escape from insect pest damage; once irrigation to avoid prolonged mid-season drought; planting castor or sunflower plants as trap crop for egg laying and destroying eggs or first stage larvae. Also, mechanical techniques are: installation of light traps; mechanical weeding at 15-20 days after sowing; collection of egg masses or early instars larvae from trap crops. *S. litura* can be controlled biologically by *Telenomus remus* (Nixon), *Apanteles africanus* (Cameron), *T. chilonis* and *B. hebetor* or by spraying SNPV, B.t., insect pathogenic fungus *Nomuraea rileyi* (Farlow). B.t is occasionally used to manage early instar lepidopteran larvae pests. If the insect population crosses ETL applying insecticides are advised. Methyl parathion 2% dust, Monochrotophos 36 SL, Quinolphos 25 EC, Endosulfan 35 EC, and Trizophos 40EC are effective against tobacco bud worm. Poison baits with Monocrotophos 36 SL or Carbaryl, rice bran, jaggery and water can use to control the grown up larvae (Jasani, 2009; Mossler & Aerts, 2009). Also, Sahayaraj & Martin (2003) stated that *Rhynocoris marginatus* (F.) significantly reduced *S.litura* population (85.89%).

3.13 *Heliothis armigera* (Hübner) (Lepidoptera: Noctuidea)

Identification: refer to the section on identification of *H. armigera* for cotton.

Life cycle and damage: it is known as gram pod borer. Female lays the eggs singly on young leaves and flower buds. Larvae feed on the foliage, prefers flowers and buds. When tender leaf buds are eaten symmetrical holes or cuttings can be seen upon unfolding of leaflets. Pupate is formed in the soil (Jasani, 2009).

Monitoring: sex pheromone traps can use for monitoring adult male population (Jasani, 2009).

Economic and action thresholds: economic threshold is the presence of two eggs or one larva per plant (Chaturvedi, 2007).

Control: ploughing deeply in summer; releasing *T. chilonis*, *Chrysoperla carnea* (Stephens), and the reduviid predator *R. marginatus* are effective for protecting groundnut from pest infestations. Application of safe chemical insecticides is recommended only if the insect population crosses the ETL. Endosulfan, Monocrotophos, Quinolphos, or Chloropyriphos is found effective (Jasani, 2009; Sahayaraj & Martin, 2003).

3.14 *Elasmopalpus lignosellus* (Zeller) (Lepidoptera: Pyralidae)

Identification: The lesser cornstalk borer is small and slender with alternating green and brown bands on its body. The eggs are oval about 0.6 mm long and 0.4 mm wide. When first deposited, they are greenish, soon turning pinkish, and eventually reddish. The larvae have brown and blue (or tan) alternating rings down the body. The pupae are yellowish initially turning brown and then almost black just before adults emerge. Pupae are about 8 mm long and 2 mm wide and the cocoons measure about 16 mm in length and 6 mm in width (Gill et al., 2008; Mossler & Aerts, 2009).

Life cycle and damage: The female deposits nearly all her eggs below the soil surface adjacent to plants. The eggs hatch in two to three days. The larvae live in the soil, constructing tunnels from soil and excrement tightly woven together with silk. Lesser cornstalk borers frequently attack seedlings emerging from the ground or shortly thereafter. It is an erratic pest with outbreaks and plant damage usually occurring during dry periods on sandy soils. The larvae also can be a severe problem from pegging time until harvest. They construct a cocoon of sand attached to the pod or stem at the point of penetration. Normally there are six instars, but the number of instars can range from five to nine depending on environmental conditions. At larval maturity, caterpillars construct pupal cells of sand and silk at the end of the tunnels. Pupa develops about 9 to 10 days, with a range of 7 to 13 days. *E. lignosellus* overwinter as larvae or pupae in the soil. This species has three or four generations annually (Gill et al., 2008; Mossler & Aerts, 2009).

Monitoring: monitor stems and lateral limbs for webbing and live larvae. For scouting the pest, monitor higher, drier parts of the field, plants on the end of rows, and those without adjacent plants. Also, determine the percentage of damage to the crop (Linker et al., n.d.).

Economic and action thresholds: economic threshold is when 10% of the monitored samples are affected (Linker et al., n.d.).

Control: Outbreaks occur during periods of hot and dry weather. Rainfall or irrigation will greatly reduce the threat of damage. Keeping the land free of weeds and grass for several weeks before planting; and earlier planting may prevent late season losses. Liquid insecticides directed at the base of host plants or granules applied to the soil can be effective but hot, dry conditions often reduce the longevity of registered insecticides (Gill et al., 2008).

Groundnuts are stored both as unshelled pods and as kernels for different uses. Some of the groundnut pests may infest kernels in storages. The ones which attack whole kernels in storages usually develop and feed inside the kernels of grain. These pests are not usually capable of existence outside the grain kernel as immature insects. Examples of whole grain pests are the rice weevil, *Sitophilus oryzae* (L.); the granary weevil, *Sitophilus granarius* (L.); and the lesser grain borer, *Rhyzopertha dominica* (F.). Other insect pests which attack stored groundnut are usually unable to penetrate whole grain. These insect pests however, can attack grain after it has been either mechanically broken or attacked by whole grain insects. Examples of these secondary pests are the confused flour beetle, *Tribolium confusum* (Jacquelin du Val.); red flour beetles, *Tribolium castaneum* (Herbst); Indian meal moth, *Plodia interpunctella* (Hubner); almond moth, *Ephestia cautella* (Walker); and the sawtoothed grain beetle, *Oryzaephilus surinamensis* (L.) (Fig. 3) (Koehler, n.d.). *Caryedon serratus* (Olivier) is the only major pest of groundnut that infests unshelled nuts (Ranga Rao et al., 2010).

Monitoring: monitoring pests in silos and storages should be done once in fortnight; so remedial measures can be taken as soon as infestation is noticed. Insect traps (sticky traps, light traps, pitfall traps, and pheromones) are effective in detecting insects which are placed either indoors or outdoors. However, estimating insect's population is difficult from a trap because some of the insects are inside the kernels and cannot be trapped. So, the symptoms of damage should be considered too (Koehler, n.d.; Ranga Rao et al., 2010).

Table : Economically important insect pests of groundnut in storage.	
Latin name	Common name
Abasverus advena (Waltl.)	Foreign grain beetle
Alphitobius diaperinus (Panzer)	Lesser mealworm
Araecerus fasciculatus (De Geer)	Coffee bean beetle
Attagenus megatoma (L.)	Black carpet beetle
Carpophilus dimidiatus (F.)	Corn sap beetle
Caryedon serratus (Olivier)	Groundnut bruchid
Corcyra cephalonica (Stainton)	Rice moth
Cryptolestes pusillus (Schoenherr)	Flat grain beetle
Dermestes lardarius (Linnaeus)	Carpet beetle
Elasmolomus sordidus (F.)	Pod sucking bug
Ephestia cautella (Walker)	Almond moth
Lasioderma serricorne (F.)	Cigarette beetle
Latheticus oryzae Waterhouse	Longheaded flour beetle
Liposcelis sp.	Booklouse – several species
Necrobia rufipes (De Geer)	Checkered beetle
Oryzaephilus mercator (Fauvel)	Merchant grain beetle
Oryzaephilus surinamensis (L.)	Saw-toothed grain beetle
Plodia interpunctella (Hubner)	Indian meal moth
Sitophilus oryzae (L.)	Rice weevil
Stegobium paniceum (L.)	Drugstore beetle
Tenebrio molitor L.	Yellow mealworm
Tenebroides mauritanicus (L.)	Cadelle
Tribolium castaneum (Herbst)	Red flour beetle
Tribolium confusum Jacquelin du Val	Confused flour beetle
Trogoderma sp.	Khapra beetle
Trogoderma granarium Everts	Khapra beetle
Trogoderma inclusum LeConte	Larger cabinet beetle
Typhea stercorea (L.)	Hairy fungus beetle

Fig. 3. Stored product insect pests of ground nut (Ranga Rao et al., 2010).

Control: sanitize and clean up the empty storage bins by spraying Malathion, or Cyfluthrin thoroughly of waste materials such as old grain, trash, or feed sacks that may furnish living quarters for insects. Clean dry grain may be protected with Pirimiphos-methyl or (S)-Methoprene (DIACON II). Grain treated with protectants should be inspected at monthly intervals to guard against the possibility of infestation. Also, after storing the grain the surface would be treated with Diatomaceous Earth, Diacon II, Pirimiphos-methyl, or Dipel Dust. Fumigation of stored grain with different insecticides such as Detia, Fumitoxin, Gastion, Gastoxin, Phostek, Phostoxin, Quick Phos, carbon dioxide, chloropicrin, or magnesium phosphide is recommended if the entire mass of the product is infested (Koehler, n.d.; Ranga Rao et al., 2010).

Caryedon serratus is probably the major pests of stored groundnuts introduced here:

3.15 *Caryedon serratus* (Oliver) (Coleoptera: Bruchidae)

Identification: Groundnut borer adult is 4 to 7 mm long, with small black markings on the elytra. It is readily distinguished from other pests of groundnuts by its very broad hind femur, serrate antennae and elytra that do not completely cover the abdomen (Ranga Rao et al., 2010).

Life cycle and damage: It is the only species that can penetrate intact pods to infest the kernels. Infestation of the harvested groundnuts can occur while the crop is being dried in the field, stored near infested stocks or crop residues. Females attach their eggs to the outside of pods or kernels and incubation period is about 4 to 6 days. The first instar larva hatches and burrows directly through the pod wall to reach the kernel, where the larva feeds and develops. Each larva feeds solely within a single kernel. A single larva can make a large excavation in the cotyledons, but no sign of damage is visible externally at this stage. Mature larvae emerge partially or completely from the pod and construct an oval papery cocoon. The life cycle of *C. serratus* is about 60 days under optimum conditions of 30°C and 70% r.h.

3.16 *Bandicota bengalensis* (Kok.); *Rattus meltada pallidior* (gray); *Tatera indica* (Hardwicke)

Among the oil seed crops groundnut often suffers severe attack by rodents. Rodents damage both the standing crop and stored products. They damage the whole or the branches of the plant during burrowing and remove the pods at the mature and harvesting stages and take them into their burrows. *B. bengalensis* is the most predominant and widespread pest of agriculture in wet and irrigated soils. In dry land *T. indica* and then *R. meltada* are the predominant rodent pests. The chronic damage ranging from 2% to 15% persists throughout the country and severe damage, sometimes even up to 100% loss of the field crop (Ghewande & Nandagop, 1997).

Control: clean cultivation; proper soil tillage; crop scheduling; barriers; repellents and proofing are cultural techniques which may reduce rodent damages. Control weeds, which are an important component of the rodents diet. Apply rodenticides like poison baiting of rodents with zinc phosphide; and fumigation with aluminium phosphide are common in agricultural fields (Parshad, 1999).

3.17 Crows (*Corvus splendens* Viellot), Pigeons (*Columba livia* Gmelin) and Black ibis (*Psaubis papillasa* Temminck)

Birds are relatively less important as pests of groundnut except pigeons and crows, which eat the freshly sown seeds leading to large gaps in the field (Ghewande & Nandagop, 1997). Brooks et al., (1988) assess 164 fields selected along road transects in Pakistan for vertebrate pests loss of groundnut. Lesser bandicoot rat (*Bandicota bengalensis* Gray) and the short-tailed mole rat (*Nesokia indica* Gray) were the most important. They often remove groundnut pods below ground without killing or otherwise damaging the plants. Also, the wild boar (*Sus scrofa* L.), desert hares (*Lepus nigricollis* Blanford); crested porcupines (*Hystrix indica* Kerr) and house crows (*Corvus splendens* Vieillot) cause damage and loss to the crop.

4. Soybean

Glycine max (L.) Merrill: known as Soya bean or soybean; about 35% of the world's production is in the United States, 27% Brazil, 19% Argentina, 6% China and 4% in India (Wikipedia, 2011e).

4.1 *Tetranychus urticae* (Koch) (Acari: Tetranychidae)

Identification: refer to the identification section of *T. uricae* in cottonseed.

Life cycle and damage: The mites generally over-winter as adult females in sheltered areas, such as plant debris and field margins. As the weather turns warm, mites become active in search of food and egg laying sites. Spider mites disperse by crawling, so infestations tend to spread slowly from field edges. *T. uricae* can complete up to seven generations in soybean per year (with generation development overlapping). Mites feed on individual plant cell contents on the underside of leaves. Each feeding site causes a stipple and severe stippling causes yellowing, curling and bronzing of the leaves. Eventually, the leaf will dry up and fall off (OMAFRA, 2011).

Monitoring: monitoring should be done weekly with shaking leaves onto a white piece of paper to see the actual mites moving around (OMAFRA, 2011).

Economic and action thresholds: action threshold is when four or more mites is observed per leaflet or one severely damaged leaf per plant prior to pod fill (OMAFRA, 2011).

Control: drought-tolerant varieties will minimize the effect of *T. uricae*. If number of mites exceed the action threshold, application of an acaricide may be necessary (OMAFRA, 2011).

4.2 *Thrips palmi* (Karny) and *Frankliniella schultzei* (Trybom) (Thysanoptera: Thripidae)

Identification: for identification *T. palmi* refer to the cotton. *F. schultzei* adults are brown, although some specimens are yellow with eight antennal segments; ocellar setae III located between the posterior hind ocella; ostocular seta I present; pronotum with posteroangular setae are slightly longer than posteromarginal setae; campaniform sensilla absent; tergite VIII with posteromarginal comb of microtrichia weakly developed (Viteri et al., 2010).

Life cycle and damage: Females of *F. schultzei* insert their eggs in flower tissue so they are known as flower thrips. Nymphs and adults feed in growing points and inside flowers. Thrips damage can result in flower abortion and pod distortion. It has two larval instars and two inactive and non-feeding stages (Bailey, 2007). In drier conditions *F. schultzei* species were collected in higher proportion on leaves and flowers. *T. palmi* can transmit tospovirus and *F. schultzei* transmit tobacco streak virus (TSV) and tobacco ringspot virus (TRSV) to soybean.

Monitoring: monitoring should be done from emergence until plants have six to eight true leaves. Shake plants on the soil or cloth and count the number of adults and larvae. At least 20–30 plants should be sampled randomly from across the field (Bailey, 2007).

Economic and action thresholds: action threshold is observation of more than four to six thrips per flower (Bailey, 2007).

Control: thrips can be controlled biologically by pirate bugs, lacewing larvae, ladybirds, *Amblyseius cucumeris* (Oudemans) and *Amblyseius swirskii* (Athias-Henriot). If the threshold is exceeded application of systemic pesticides such as Dimethoate can reduce thrips population (Bailey, 2007).

4.3 *Aphis glycines* (Matsumura) (Homoptera: Aphididae)

Identification: The soybean aphids are small, pale yellow with black cornicles and a pale yellow tail. Adults may be winged or wingless. Nymphs are smaller than adults and wingless. Eggs are small, spherical and yellow when first laid but turn a dark brown.

Life cycle and damage: females usually lay eggs along the seams of the buckthorn bud. In the spring, nymphs hatch and aphids undergo two generations as wingless females on the buckthorn. The third generation develops into winged adults that migrate to soybean plants. The aphids then continue to produce wingless generations until the soybean plants become crowded with aphids and the plants experience a reduction in quality. Then, winged forms are produced to disperse to less-crowded soybean plants. There can be as many as 18 generations of aphids per year on soybeans. Males are only born in the fall so that the females and males can mate to produce the egg on buckthorn. Aphids suck the sap from leaves and stems of the crop. At high populations, aphids can cause the plants to abort flowers, become stunted, reducing pod and seed production and quality. Aphids also excrete honeydew, which can act as a substrate for grey sooty mould development. The soybean aphid may also be a vector of soybean mosaic virus (OMAFRA, 2011).

Monitoring: Monitoring should be done until the soybean is well into the R6 stage. Scout weekly; or as if the populations approach the threshold monitor more frequently (every 3-4 days). Monitor 20-30 random plants across the field and estimate the number of aphids per plant in that field (OMAFRA, 2011).

Economic and action thresholds: The threshold is 250 aphids per plant and actively increasing on 80% of the plants from R1 up to the R5 stage of soybeans (OMAFRA, 2011).

Control: some of the soybean cultivars such as 'Dowling', 'Jackson', and 'Palmetto', showed resistance to aphids. Insidious flower bug, (*Orius insidiosus* (Say)), twospotted lady beetle (*Adalia bipunctata* L.), sevenspotted lady beetle (*Coccinella septempunctata* L.), the spotted lady beetle (*Coleomegilla maculata* De Geer), the polished lady beetle (*Cycloneda munda* (Say)), the multicolored Asian lady beetle (*Harmonia axyridis* (Pallas)), the convergent lady beetle (*Hippodamia convergens* Guérin-Méneville), and the thirteen spotted lady beetle (*Hippodamia tredecimpunctata* L.) can biologically control the aphids. Also, pathogenic fungi including *Pandora neoaphidis* (Remaud. et Henn.) Humber, *Conidiobolus thromboides* Drechsler, *Entomophthora chromaphidis* Burger et Swain, *Pandora* sp., *Zoophthora occidentalis* (Thaxter) Batko, *Neozygites fresenii* (Now.) Remaud. et Keller, and *Lecanicillium lecanii*, (Zimm.) Gams et Zare cause infection in an outbreak aphid population (Nielsen & Hajek, 2005).

4.4 *Bemisia tabaci* biotype B (Genn.) (Hemiptera: Aleyrodidae)

Identification: refer to the identification of *B. tabaci* in cotton section.

Life cycle and damage: The B-biotype is a pesticide-resistant strain of *B.* tabaci. The female lays eggs on younger leaves. Nymphal activity is further down the plant about 5-7 nodes below the plant top. All stages (unlike eggs) secrete large amounts of sticky honeydew and sooty mould develops on honeydew reduces photosynthesis of the leaves. Severe infestations in young plants can stunt plant growth and greatly reduce a crop's yield potential. Later infestations can reduce the number of pods set, seed size, and seed size uniformity, thus reducing yield and quality. However, pod and seed discolouration are a major marketing problem where pods are picked green (Brier, 2011).

Monitoring: monitor eggs and young nymphs from uppermost and youngest leaves; and older nymphs and pupae from older leaves. Yellow sticky traps can use for sampling adults. The presence of honeydew and sooty mould may also indicate *B. tabaci* attack.

Economic and action thresholds: No action threshold has been set for *B. tabaci* on soybean.

Control: avoid successive plantings of summer pulses to prevent movement from early to late crops. Control weed hosts such as rattlepod and milk thistle; and irrigate crops to reduce moisture stress. Also, overhead irrigation washes honeydew off leaves, lessening the risk of sooty mould. Nymphs are biologically controlled by *Encarsia* sp. and *E. mundus*. No pesticides are specifically registered for controlling *B. tabaci* (Brier, 2011).

4.5 *Nezara viridula* (Linnaeus), *Euschistus servus* (Say) and *Acrosternum hilare* (Say) (Hemiptera: Pentatomidae)

Identification: they are known as Southern green stink bug (SGSB), brown stink bug (BSB) and green stink bug (GSB), respectively. All stink bug adults are shield-shaped. Green and southern green stink bugs are bright green and measure 14-19 mm long. The major body regions of the GSB are bordered by a narrow, orange-yellow line. BSB is dull brownish-yellow in color and 12-15 mm long. At first eggs are barrel-shaped and for the SGSB; they are creamy, cylindrical eggs of the measure 1 by 0.75 mm and develop a pinkish hue before hatching. The eggs of BSB are white, kettle-shaped, and slightly smaller than those of the green stink bug. Eggs of the green stink bug are yellow to green, later turning pink to gray and about 1.4 x 1.2 mm. The nymphs of all three species are smaller than adults, but similar in shape. Nymphs of the BSB and SGSB are light green. For SGSB, however, nymphs have two series of white spots along their backs. GSB nymphs are black when small, but as they mature, they become green with orange and black markings. The nymphs remain clustered near the remains of the eggs (Bijlmakers, 2008; Gomez & Mizell, 2008). *A. hilare* can be differentiated from the *Nezara* spp. by its elongated ventral ostiolar canal (external outflow pathway of metathoracic scent gland), which extends well beyond the middle of its supporting plate, while that of *N. viridula* is shorter and does not reach the middle of supporting plate. *A. hilare* has black outermost three antennal segments (Wikipedia, 2011c).

Life cycle and damage: stink bugs overwinter as adults and they become active during the first warm days of spring when temperature rises above 21°C. Adults are strong fliers and will readily move between weeds and other alternate hosts where they complete their first generation. They are attracted most often to plants with growing shoots and developing seeds or fruits. Stink bugs tend to move from host to host as peak reproduction of earlier hosts passes and that of their plants approaches. Both nymphs and adults attack primarily the seeds and pods of soybean. The degree of damage depends to some extent on the developmental stage of the crop. They also will feed on soybean stems (Olson et al., 2011). Female of the *N. viridula* lays about 300 eggs on the underside of the leaves and stuck them together in batches of 50-60. After the eggs hatch, nymphs disperse and start feeding. However, the first instar nymphs don't feed. SGSB is the major pod-sucking bug in soybeans due to its abundance, widespread distribution, rate of damage and rate of reproduction. Adults prefer pods with well-developed seeds and nymphs are unable to complete their development prior to pod-fill. They can damage soybeans until pods are too hard (i.e. very close to harvest). Damage to young pods cause deformed and shrivelled seeds and reduce yield. Damaged seeds in older pods are blemished and difficult to grade out, which reduce harvested seed quality. *N. viridula* also damages buds and flowers (Bijlmakers, 2008; Brier, 2011). In *E. servus* reproductive diapauses, they color change from green to reddish-brown. Reddish-brown insects lived longer than green ones and laid no eggs. However, green females lay about 18 egg masses; after approximately four to five weeks eggs hatch and form the nymphs. The nymphs have five instars and after about 29 days they develop to the adult. Pod is pierced or kinked and seed is dimpled or blemished. *E. servus* has as many as four to five or more generations per year in Florida but only two generations a year in southeastern USA. However, *N. viridula* can stay in peanut for their third generation, or leave maize for full-season soybean or cotton where they produce their third or even fourth generation. Both SGSB and BSB preferred soybean significantly more often than B.t. cotton, non-B.t. cotton, and peanut (Gomez & Mizell, 2008; Olson et al., 2011).

Monitoring: monitor twice weekly from flowering until close to harvest. Beat sheet sampling is the most efficient monitoring method with at least six sites for adults or 10 sites for nymphs. The standard sample unit consists of five one-metre non consecutive lengths of row within a 20 m radius (Brier, 2011). Sweep net and ground cloth procedures are common sampling methods. Place one four-foot cloth per location and record the number per four-foot sample. In the case of sweep net; use a 15 sweep net and take a series of twenty five sweeps at each location (Johnson et al., 2009). Stink bug trap with the aggregation pheromone is also way of monitoring and capturing the bugs (Gomez & Mizell, 2008).

Economic and action thresholds: economic thresholds depends on the crop size (seeds per m^2) and when bugs first infest a crop. The maximum bug damage permitted being only 2% or for *N. viridula* when the adults number ranges from 0.3-0.8/m^2 (Brier, 2011).

Control: control during early pod-fill before nymphs reach a damaging size. Avoid sequential plantings of summer legumes as bugs population will move from earlier to later plantings. Avoid cultivar and planting time combinations that are more likely to lengthen the duration of flowering and podding. *Trissolcus basalis* (Wollaston), and *Trichopoda giacomellii* (Blanchard) may control stink bugs (Brier, 2011). *A. hilare* is parasitized by the tachinid fly, *Trichopoda pennipes* (F.). The pheromone methyl (E,Z,Z)-2,4,6-decatrienoate may be used to attract the bugs away from fields (Wikipedia, 2011c). Also, trap crops baited with pheromone to catch females for destruction, or even a pheromone-based disruption of orientation behavior to decrease the mating success, are possible semiochemical techniques to suppress populations of *E. servus* second generation (Borges et al., 2001).

4.6 *Piezodorus oceanicus* (Montrouzier) (Hemiptera: Pentatomidae)

Identification: The red banded shield bugs are similar in shape to *N. viridula* but smaller, paler and with pink, white or yellow bands. Eggs ringed by small spines and have a central pale band. Newly hatched nymphs are orange with black markings. Larger nymphs are pale green with dark red and brown in the centre of their back and late nymphs may turn to the pale pinkish brown. It was previously classified as *Piezodorus hybneri* (Gmelin), and recently as *P. grossi* (Ahmad) or *P. grossi* (Staddon) (Bailey, 2007; Brier, 2011).

Life cycle and damage: Egg rafts contain 15-40 eggs and after 4-5 days hatch. *P. oceanicus* has five nymphal stages. Nymphs usually reach a damaging size during mid to late pod-fill. They damage is similar to *N. viridula*, with early damage reducing yields, and later damage reducing the quality of harvested seeds. *P. oceanicus* generally has four generations per year (Bailey, 2007; Brier, 2011).

Monitoring: Look for the distinctive twin-row egg rafts which indicate the presence of *P. oceanicus*. Beat sheet sampling is the efficient method for monitoring bugs (Brier, 2011).

Economic and action thresholds: no thresholds have been set for red banded shield bug.

Control: Eggs may be parasitised by *T. basalis*. No insecticides are specifically registered against *P. oceanicus* (Brier, 2011). Brier & McLennan (2006) stated that the addition of 0.5% salt (NaCl) to Decis (deltamethrin) will have a greater impact on *P. oceanicus* population. Also, high yield of the crops contributed to the lower percentage of damaged seeds.

4.7 *Oberea brevis* (Swederus) (Coleoptera: Lamiidae)

Identification: girdle beetles are yellow at first then change to red and brown on the head, thorax and bases of elytra. Larvae are white, soft-bodied with a dark head (TNAU, n.d.-c).

Life cycle and damage: The female deposits a single egg in the petiole of a leaf near the top of the plant. The larva feeds into the petiole and on down into the stem, bore the inside of the stem so it is a stem borer pest. Only one larva is present in each infested plant. The full-grown larva prepares a fibrous cocoon and pupates within the girdled portion of the stem. The adult emerges from the girdled section in 8-11 days. However, most larvae enter diapauses within the cocoon (Kapoor et al., 1972). The leaves of infected plant are unable to get the nutrient and thus are dried up. In later stages the plant is cut at about 15-25 cm above the ground (TNAU, n.d.-c). The weight of the grain from the top-cut portions of infested plants was significantly reduced in relation to that of healthy plants and their germination capacity was also seriously affected (Gangrade & Singh, 1975).

Monitoring: beetles may be collected in sweep net samples. Thereafter, look for lodged plants, which may indicate the presence of stem borer infestations. Cut, open the stem and look for the presence of the larva and its feeding damage. Also, count number of larvae and lodged plants (Boyd & Bailey, 2005).

Economic and action thresholds: Out of thirty-five genotypes studied, Awasthi et al., (2003) recorded stem tunneling (%) from 6.46 to 32.16% and the economic injury level to be 26%. In another study, Rao et al., (2007) declared 5% incidence of *O. brevis* for economic threshold.

Control: summer ploughing; avoid intercropping with maize or sorghum; crop rotation; avoid excess nitrogenous fertilizers; collecting and destroying infested plant parts and egg masses can decrease population of beetles. If the threshold is exceeded application of insecticide is recommended. Apply phorate 10 kg/ha or carbofuran 30 kg/ha at the time of sowing. Spray Endosulfan 35 EC, Dimethoate 30 EC, Quinalphos 25 EC, methyl demeton 25

EC, Monocrotophos 36 SC, Monocrolophos 36 WSC, Quinalphos 25 EC or Triazophos 40 EC at the crop age of 30-35 days and repeal after 15-20 days. Also, acephate 0.10% is effective for controlling girdle beetles from infestations (Pande et al., 2000; TNAU, n.d.-c).

4.8 *Cerotoma trifurcata* (Forster) (Coleoptera: Chrysomelidae)

Identification: The bean leaf beetle adult is 5 mm long, with or without four black spots (parallelogram shaped) on the wing covers. Adults can vary in color but are most often yellow-green, tan or red. A small, black triangle is visible at the base of the wing covers the prothorax - behind the head. The margins of the wing covers have a black border. Eggs are orange colored and lemon shaped (OMAFRA, 2011).

Life cycle and damage: Females lay eggs in small clusters in the soil at the base of the soybean plants. Newly hatched larvae feed on roots and other underground plant parts for about 30 days before pupating. Adults second generation emerges, feeds on the pods until the plants senesce. They then migrate to alfalfa fields if available or move to their over wintering sites. *C. trifurcate* overwinters in the adult stage in woodlots, leaf litter and soil debris. Defoliation injury by this pest is generally not serious and the exception is damage caused by over wintering adults to young soybean plants. Adult feeding appears as small round holes between the major leaflet veins. Cotyledons and seedling plants can be clipped off by heavier populations. They also feed on the surface of the pod, leaving only a thin film of tissue to protect the seeds within the pod. These pod lesions increase the pod's susceptibility to secondary pod diseases such as alternaria. The bean leaf beetle is a vector of bean pod mottle virus (BPMV) which causes the plant and seed to become wrinkled and mottled, reducing the quality of the seed (OMAFRA, 2011).

Monitoring: monitor from seedling stage to dry edible bean pod-fill stages. Fumigation cage, sweep net, pitfall trap and shaking over the ground cloth are the methods used for sampling. For sweep net sampling take 20-sweep samples at each sample location. Sweep as you walk down the row and calculate an average number of beetles per sweep. In the case of shaking over the ground cloth; place a two foot wide strip of cloth on the ground between the rows and shake the plants over them. Then quickly count the number of beetles. In the "seedling stage" see underside of the leaves and calculate the average number of beetles per meter of row; in the soybean "R5-R6 Stage", assess the number of pods with feeding injury or clipping; in the "prior to the dry edible bean pod-fill Stages" determine the percent defoliation that has occurred; and in the case of "dry edible bean pod-fill Stages" determine the percent of pods with feeding damage. For pod sampling collect five plant samples from each sampling location and calculate average number of pods with holes in the pod wall per five plants (OMAFRA, 2011).

Economic and action thresholds: Thresholds for bean leaf beetle are 16 adult beetles per foot of row in early seedling stages. In V3-R4 stage of soybean; the economic threshold is when defoliation exceeds 30% and for R5-R6 stage if 10% of the pods on the plants have feeding injury and the beetles are still active in the field (OMAFRA, 2011).

Control: Seed treatment using thimethoxam and late planting date, may result in low initial population of beetles. Tachinid species, *Calatoria diabroticae* (Shimer) and *Medina* sp. can control *C. trifurcata*. Application of foliar insecticide between emergence and first trifoliate using pyrethroid has been shown to reduce beetles incidence, by protecting soybean from overwintering beetles. Additional application of foliar insecticide (around blooming) was shown to further suppress of population. A mid-season foliar insecticide application is aimed at controlling the first generation of bean leaf beetle (Bennett et al.).

4.9 *Epilachna varivestis* (Mulsant) (Coleoptera: Coccinellidae)

Identification: Mexican bean beetles are copper to yellow, rounded with 16 black spots on their backs. Males can be distinguished from females by having a small notch on the ventral side of the last abdominal segment. The eggs are yellow about 1.3 mm long and elliptical. Larvae are yellow, oval, soft-bodied, grub like with darker, branched spines. The pupae are yellow and about 6 mm long. It moves very little, has fewer spines than the larva, and is most commonly found on the lower half of the soybean plant (Sanchez-Arroyo, 2009).

Life cycle and damage: females lay eggs in clusters of 40-75 on the undersides of bean leaves. They hatch in a week during warm weather or after at least two weeks under unfavorable conditions. The larvae are gregarious and feed voraciously for two to five weeks. The larva has five instars and attaches itself by the posterior end of the body to the underside of leaves, stems, or pods of the bean plants and often to parts of nearby plants and pupates. The pupal stage lasts 5-10 days. Adults and larvae feed on the underside of leaves and eat the softer leaf tissues. Damaged leaves appear netlike (Sanchez-Arroyo, 2009).

Monitoring: monitoring should be done weekly with directly observing the plant during early stages of growth. As the plant growth use sweep net or shaking the plants over the ground cloth for sampling *E. varivestis*. After shaking the plants quickly count number of the present adults and larvae (Sanchez-Arroyo, 2009).

Economic and action thresholds: economic threshold level ranges from 1 to 1.5 larvae per plant on beans or when 30–35% defoliation is observed prior to full bloom and 15% during pod-set and podfill (Sanchez-Arroyo, 2009).

Control: destruction of overwintering locations; late planting of the soybean crop; planting trap crops and resistant varieties are techniques that may reduce population build up. Releasing biological control agents such as tachinid fly, *Paradexodes epilachnae* (Aldrich) and the eulophid wasp, *Pediobius foveolatus* (Crawford) can control the number of *E. varivestis* beetles. If the threshold is exceeded application of systemic insecticides at planting has become a standard practice (Sanchez-Arroyo, 2009).

4.10 *Melanagromyza sojae* (Zehntner) (Diptera: Agromyzidae)

Identification: adults of stem fly are shining black and about 2 mm long. The larva which is named maggot is white in color and remains inside the stem. The egg measures about 0.34 mm in length and 0.15 mm in width. It is whitish, partly transparent and the yolk occupies one third of the central part of the egg (TNAU, n.d.-c).

Life cycle and damage: The females lay eggs on or in the leaves. Maggots bore the nearest vein of the leaf and subsequently reach the stem through petiole and bore down the stem. If the infected stem is opened by splitting, zigzag reddish tunnel can be seen with maggot or pupae inside it. The maggots feed on stem cortical, may extend to tap root, killing of the plant. The third larval instar, mine a hole to the epidermis to assist in the emergence of the adult and pupates in the stem. The pupal period lasted 6-12 days (TNAU, n.d.-c). *M. sojae* has four or five generations per year and pupae overwinter in the stem. The infestation significantly reduced the plant height, number of branches per plant, number of trifoliate leaves, leaf area per plant and dry matter accumulation. The infestation rate ranges from 85 to 100% due to the planting dates and being higher on the late sown crop (Savajji, 2006).

Monitoring: Stem flies reached their peak in the 5th–8th weeks after planting and declined towards the end of the season. Observe percent seedling mortality, percent stem tunneling and areas of poor stand. Also, dig up some seed and examine for the presence of insects and or damage (Savajji, 2006).

Economic and action thresholds: economic threshold level is 5% plant infestation (Rao et al., 2007). However, according to (Johnson et al., 2009) there is no rescue treatment for this pest. Estimate area and amount of stand reduction for reporting in Comments. This information will be useful when making a decision on replanting.

Control: deep summer ploughing; proper crop rotation with dissimilar crops; removing and destroying the damaged plant parts can reduce the population. However, if the threshold is exceeded application of insecticides is essential. Spray monocrotophos 36 WSC twice, at the crop age of one and three weeks. In case of severe infestation, apply phorate or lindane in the soil 10 kg/ha before sowing. Soil application of phorate 10 G, 10 kg/ha or carbofuran 3 G, 30 kg/ha at the time of sowing will prevent early infestation by stem fly. One or two sprays of endosulfan 35 EC or dimethoate 30 EC or quinalphos 25 EC can stop the damage (Pande et al., 2000; TNAU, n.d.-c).

4.11 *Asphondylia yushimai* sp. n. (Diptera: Cecidomyiidae)

Identification: Soybean pod gall midge is described from Japan and at least five named and 13 undetermined species of *Asphondylia* are found in Japan. Most of the undescribed species are morphologically very similar to one another. Yukawa et al., (2003) studied the morphological characteristics of soybean pod gall midge and DNA sequencing of several *Asphondylia* gall midges. They prove that *A. yushimai* is a distinct species of the genus *Asphondylia*, identify its winter host, and outline host alternation by this gall midge.

Life cycle and damage: females lay eggs inside young pods of soybean and cause yield losses with malformation of pods by larval infestations. Soybean pod gall midge is one of the major pests of soybean in Japan, Indonesia and China. In summer and autumn, the soybean pod gall midge has two or more generations in the pods of soybean, *G. max* or wild fabaceous and caesalpiniaceous plants (Yukawa et al., 2003). *Asphondylia* sp. cannot overwinter on soybean because the host dies back in winter; whether the soybean species overwinters on another legume or host (Gagne & Woods, 1988).

Monitoring: collect the galls and count number of larvae. Color traps present a potential tactic for monitoring soybean pod gall midge or mating disruption (RADA, 2010).

Economic and action thresholds: No action threshold has been set for *A. yushimai*.

Control: Controlling weeds beneath or around the plants and removing fallen fruit from fields (bury at least 15 cm deep or burn them) may reduce the infestation of gall midge. If a field is of no further economic use, destroy completely or remove all reproductive structures at least twice weekly (RADA, 2010). Contact county Cooperative Service Extension agent or office for latest applicable insecticides.

4.12 *Spodoptera litura* (F.) (Lepidoptera: Noctuidae)

Identification: refer to the identification of *S. litura* in groundnut. *S. litura* and *S. littoralis* are morphologically similar and distinguish with close examination of the genitalia.

Life cycle and damage: females of common cut worm lay their eggs in hair-covered clusters. Newly hatched larvae are very susceptible to dry heat, usually staying on leaves lower surfaces during the day and feeding at night. The larvae damage soybean extensively by skelatalization of leaves in early stage, severe defoliation in later stage and thus reducing the photosynthetic capacity of plants. After damaging the leaves, they also start feeding on young parts, consequently damaging 30 to 50% of the pods. It has six larvae instars. Mature caterpillars pupate in earthen cells in the soil. Occasionally this insect also cuts plant stems at the soil line and feeds directly on the pods. In severe infestations, leaves are completely gone causing the plant to die (Parker et al., 2001; Santhosh, 2008).

Monitoring: monitor *S. littora* males peak with installation of pheromone traps. Look carefully at the upper and lower surfaces of the leaves to detect the presence of larvae.

Economic and action thresholds: the economic threshold is when defoliation is at or will reach 40% before R1 stage of the crop.

Control: using pheromones to mass trapping of the males by the lure and kill technique may be applied for controlling *S. littora*. Biological control of larvae by fungal pathogen, *Nomuraea rileyi* (Samson) and B.t. can suppress population builds up (Patil & Hegde, 2010). The nematodes *S. carpocapsae*, *Neoaplectana carpocapsae*, and the fly *Exorista japonica* (Townsend) are also effective against *S. litura*. If the threshold is exceeded application of insecticides, and insect growth regulators such as neem is recommended (Santhosh, 2008).

4.13 *Heliothis armigera* (Hübner) and *Heliothis punctigera* (Wallengren) (Lepidoptera: Noctuidea)

Identification: *H. armigera* is known as America bollworm and *H. punctigera* known as Australian bollworm. For identification refer to *H. armigera* and *H. punctigera* on cotton.

Life cycle and damage: *Heliothis* spp. is one of the major pests of soybean. They attack auxiliary buds and terminals in vegetative stage and defoliate the crops. Once crops reach flowering, larvae focus on buds, flowers and pods. Young larvae are more likely to feed on vegetative terminals, young leaves and flowers before attacking pods. Crops are better able to compensate for early rather than late pod damage, however in dry land crops, significant early damage may delay podding with subsequent yield and quality losses (Brier, 2011).

Monitoring: monitor eggs and moths to determine the start of infestations and increase the chance of successful control. Monitor small larvae by opening vegetative terminals and flowers. Monitoring should be done twice weekly from early budding until late podding and sample six widely spaced locations per field. Beat sheet sampling is the preferred sampling method for medium to large size larvae (Brier, 2011).

Economic and action thresholds: Before flowering and in vegetative stage, economic threshold is when 33% leaf loss is determined or approximately 7.5 larvae/m^2 is observed. For podding soybeans currently ETL ranges from 1-2 larvae/m^2 (Brier, 2011).

Control: avoiding successive plantings of summer legumes and growing vigorous plants with adequate available moisture are cultural control techniques (Brier, 2011).

4.14 *Spilosoma obliqua* (Walker) (Lepidoptera: Arctiidae)

Identification: refer to the identification of *S. obliqua* in groundnut.

Life cycle and damage: Bihar hairy caterpillar females lay their eggs in cluster on the underside of leaves. The larva is a voracious feeder which feeds gregariously on soybean leaves, on chlorophyll mostly in early instars. In later stages the larvae eat the leaves from the margin. The leaves of the plant give an appearance of net or web and in case of severe infestation, the entire crop is damaged badly thus causing 40% defoliation of leaf area (Harish, 2008; TNAU, n.d.-c). They can severely damage mustard crops too.

Monitoring: install one light trap per hectare to catch the adults of hairy caterpillar. Also, count number of larvae in the meter rows that were selected randomly (TNAU, n.d.-c).

Economic and action thresholds: ETL is observing 5 larvae per meter row (Rao et al., 2007).

Control: summer ploughing, intercropping soybean either with pigeon pea, maize or sorghum; collect and destroying infested plant parts, egg masses and young larvae are techniques that can control *S. obliqua*. If action threshold is observed application of endosulfan 35 EC, chlorpyriphos 20 EC, and trizophos 40 EC, is advised (TNAU, n.d.-c).

4.15 *Plathypena scabra* (F.) (Lepidoptera: Noctuidae)

Identification: moths are triangular when at rest and the wingspan is about 25 mm long. The female moths have charcoal-colored wings with brown and silver patches, while wings of male moths are more uniformly charcoal in color. The male moths have larger body and eyes size than the female moths. The caterpillars of green clover worm are greenish with faint white strips along the body. Sometimes the stripes are not obvious. They have four pairs of pro legs and move with a looping motion similar to the soybean looper. When disturbed, these larvae become very active and fall to the ground (Baldwin et al., 2011).

Life cycle and damage: females lay eggs singly on the underside of soybean leaves. After 3-4 days, the eggs hatch and the larvae start to feed on the leaves. The larva has 6 instars and develops in about 14 days. Green clover worms only infest foliage. They make holes in the leaves and are damaging only under very high populations or in combination with other defoliators. Last instar larvae burrow into the soil or plant debris to pupate and emerge as adults in 7-10 days to repeat the cycle (Hadi et al., 2011).

Monitoring: use ground cloth procedure, roll the cloth beneath the canopy from one row over to the next row and shake the plants on cloths. Make one four-foot shake-cloth sample per location and record the number of larvae (Johnson et al., 2009).

Economic and action thresholds: ETL is when 5.2 larvae is observed per row-foot (17 larvae per row-meter) between bloom and pod-fill. Also, if defoliation reaches 40% pre-bloom, 20% during bloom and pod-fill, or 35% from pod-fill to harvest (Hadi et al., 2011).

Control: beneficial insects and pathogens such as the fungal pathogen *N. rileyi*, B.t, parasitoid *Cotesia marginiventris* (Cresson), and *Rogas nolophanae* (Ashmead), can control clover worms population. If the threshold is exceeded application of insecticides is advised. Azinphos-methyl, (Sniper®2-E), Carbaryl (Sevin XLR Plus), and Lorsban 4E are common chemicals labeled for green cloverworm control (Daigle et al., 1988; Gouge et al., 1999).

4.16 *Pseudoplusia includens* (Walker) (Lepidoptera: Noctuidae)

Identification: Soybean looper moth has mottled brown forewings with a bronze to golden glitter. At the center of the forewings, two prominent silver markings are usually visible. The larva is light green with several thin light lines in the body length. The body of caterpillar is largest at the rear and tapering to the head. Loopers form the characteristic hump or "loop" when crawling. They have three pairs of prolegs (Hadi & Bradshaw, 2011).

Life cycle and damage: females lay eggs on the underside of soybean leaves. After 3 days the eggs hatch and first instar larvae feed on soybean leaves. The larvae start feeding with leaves at the middle of the plants and working upward. The larvae have 6 instars. Loopers eat large holes in the leaves and, under high populations, can strip an entire field. The pupae are covered by silken cocoons attached on the underside of soybean leaves. This pest usually occurs in fields that have been previously sprayed. Soybean loopers usually reach higher populations in areas where cotton and soybeans are grown together. Their infestations is not as explosive, but they can build up to the economic threshold (Baldwin et al., 2011).

Monitoring: monitor population of moths with pheromone traps to attract, collect and count number of moths.

Economic and action thresholds: ETL is when defoliation reaches 40% pre-bloom, 20% during bloom and pod-fill, or 35% from pod-fill to harvest (Hadi & Bradshaw, 2011).

Control: Planting an early maturing soybean variety will allow the soybean crop to escape damaging populations of the loopers. Pheromone traps can use as control tools to capture insects, disrupt insect mating and as lures to attract insects to insecticidal baits. Application of foliar insecticides is advised if the threshold is exceeded (Baldwin et al., 2011).

4.17 *Anticarsia gemmatalis* (Hübner) (Lepidoptera: Noctuidae)

Identification: The moths have ash gray, light yellowish-brown, or dark reddish brown forewings and light brown hindwings with a row of light colored spots near the margin. The eggs are white, slightly oval, about 1-2 mm diameter and flattened on its lower surface. Velvetbean caterpillar is greenish, brown, or almost black caterpillars with a broad lighter band down each side. Head is prominent and yellow or orange. Caterpillars have five pairs of prolegs and the last pair looks like forked tail. When disturbed, they wiggle violently. The pupa is light green at first and turns brown then (Baldwin et al., 2011; Kathryn, 2008).

Life cycle and damage: females lay eggs on the underside of leaves or maybe on the upper surfaces of leaves, on the petioles and even on the stems. After about 3 days eggs hatch. This caterpillar eats the entire leaf and will strip whole fields when high numbers are present. They move downward and defoliate the plants if uncontrolled. It has six instar larvae and the pupae will form below the soil. The pupal stage usually lasts about 7 to 11 days due to the weather conditions. The life cycle is completed in four weeks. Pods sometimes will be attacked when defoliation is severe. Damaging populations usually limited to late-planted fields, especially double-crop soybeans (Baldwin et al., 2011; Kathryn, 2008).

Monitoring: Direct observation of caterpillars on the plant is the best method of sampling for early stages of plant growth. Sweep net, black light and pheromone traps are good indicators of moth presence. If the moths are detected, searching for eggs and larvae and counting them is the next step.

Economic and action thresholds: ETL in vegetative and flowering stage of the crop are 33% and 20%, respectively or the presence of three larvae per square meter (Bailey, 2007).

Control: planting or use of early maturing varieties of soybean, and use of trap crops are the cultural techniques appropriate for reducing pest's populations. Caterpillars can be controlled by pathogens such as NPV and B.t or the parasitoid *Winthemia rufopicta* (Bigot). Application of Dimilin at 2 ounces product per acre will provide preventive control of *A. gemmatalis*. Dimilin should be applied at the vegetative growth of soybeans and as pod formation begins (at or shortly after bloom). The probability of a velvetbean caterpillar problem is higher on later maturing soybeans (Baldwin et al., 2011; Kathryn, 2008).

5. Mustard

Brassica napus L.: known as Canola, Rapeseed, Rutabaga, and Swede.
Brassica juncea L.: known as Brown and Oriental mustards, is grown commercially in Canada, the UK, Denmark and the USA.
Brassica nigra L.: known as Black mustard, is grown in Argentina, Chile, the USA and some European countries.
Brassica alba L. or *Brassica hirta* Moench.: known as White and yellow mustards, is grown wild in North Africa, the Middle East and Mediterranean Europe.
Brassica rapa L. or *Brassica campestris* L.: known as field mustard and turnip mustard.

5.1 Snails: *Cernuella virgata* (Da Costa), *Cochlicella acuta* (Müller), *Prietocella barbara* (Linnaeus) (Eupulmonata: Hygromiidae), and *Theba pisana* (Müller) (Eupulmonata: Helicidae); Slugs: *Limax cinereoniger* (Wolf) (Eupulmonata: Limacidae) and *Deroceras reticulatum* (Müller) (Eupulmonata: Helicidae)

Identification: snails are known as common white snail, pointed snail, small pointed snail and white Italian snail, respectively. The common names of slugs are black keeled slug and reticulated slug, respectively. Shell diameter of common white snail ranges from 10 to 15 mm. The coiled white shell has a brown band around the spiral in some individuals while others completely lack this banding. The umbilicus is open and circular. Under magnification, regular straight scratches are visible across the shell. The common white snail and white Italian snail are the same just umbilicus are different and the scratching per etches on the shell. Pointed snail is fawn, grey or brown with some white markings. The shell is conical in shape. The length of the shell of a mature snail is up to 18 mm. The ratio of the shell length to its diameter at the base is always greater than two. Small pointed snail shell is fawn, grey or brown in color. The length of the shell of a mature snail is up to 10 mm. The ratio of the shell length to its diameter at the base is always two or less. The black keeled slug is longer than 20 mm and has a uniform grey to black body color. This species has a prominent ridge (keel) running along the mid-dorsal line, from the mantle to the tip of the tail. The reticulated slug is black to brown, with elongated soft and slimy body about 40-60 mm long with a ridge down its back (Bailey, 2007).

Life cycle and damage: They are hermaphrodite so both male and female can lay eggs. Eggs are laid into moist soil and hatch after 2–4 weeks. They usually become sexually mature after one year. Feeding of these pests may retard development of young crops. They destroy crops by eating roots, leaves, stems and fruits. Damage to the canola during emergence, is usually difficult to detect if seedlings are chewed to the ground. Black keeled slug feed on the soil surface, as well as below the ground where they burrow down and attack germinating seeds. Young canola seedlings are particularly vulnerable. This species is of relatively greater importance in drier areas. For reticulated slug egg clusters are laid in the top soil. After 2 weeks the eggs hatch and the juveniles feed in winter and spring and aestivate over summer to become sexually mature at one year old (Micic et al., 2007).

Monitoring: monitor snails by square sampling quadrates. Quadrates are placed on the ground, the number of snails counted and convert to snails per square meter (Bailey, 2007).

Economic and action thresholds: action thresholds is observing 5 snails per m² in emerging canola crops (Bailey, 2007). Micic et al., (2007) declared that action threshold for small pointed snail is 20 snails per m², for common white snail 5 per m², black keeled slug 1-2 per m², reticulated slug 1-2 per m² and in the case of white Italian snail it is 5 snails per m².

Control: They can be controlled by birds and lizards. A parasitic fly, *Sarcophaga penicillata* (Villeneuve) is a bio-control agent for controlling pointed snail. However, the best way is baiting but it should be done in the cool moist condition, when snails are active. As bait may have some residue contamination in kernels so it must be completed at least 8 weeks before harvest. For 5-80 snails per m² apply bait at 5 kg/ha and over 80 snails per m² apply bait at 10 kg/ha. Metaldehyde-containing baits are frequently used for snail control (Bailey, 2007).

5.2 *Lipaphis erysimi* (Kalt) (Homoptera: Aphididae)

Identification: Aphids are small, soft-bodied, pearl-shaped with a pair of cornicles projecting out from the fifth or sixth abdominal segment. Wingless female aphids are yellowish green, gray green or olive green with a white waxy bloom covering the body. The winged females have a dusky green abdomen with dark lateral stripes separating the body segments and dusky wing veins. Males are olive-green to brown in color (TIFP, n.d.-b).

Life cycle and damage: Mustard aphids infest the crops right from vegetative stage to pod stage. Both nymph and adults suck the sap from leaves, buds and pods. Curling may occur for infested leaves and at advanced stage plants may wither and die. Plants remain stunted and sooty molds grow on the honey dew excreted by the insects. The infected filed looks sickly and blighted in appearance. Rainy and humid weather help in accelerating the growth of insects. They have four nymphal instars (TIFP, n.d.-b).

Monitoring: install yellow stick trap to monitor aphid population (TIFP, n.d.-b).

Economic and action thresholds: The economic threshold level for mustard aphid for different regions of India is 23–25 aphids per plant (Chattopadhyay et al., 2005).

Control: planting tolerant or resistant varieties can reduce population build up. *Cocciniella septempunctata* (L.), *Menochilus sexmaculata* (F.), *H. variegata*, *Cheilomenes vicina* (Mulsant), *Sphaerophoria* spp., *Eristallis* spp., *Metasyrphis* spp., *Xanthogramma* spp. and *Syrphus* spp. are most efficient predators of the mustard aphid (TIFP, n.d.-b).

5.3 *Myzus persicae* (Sulzer) and *Brevicoryne brassicae* (Linnaeus) (Homoptera: Aphididae)

Identification: They are known as peach potato aphid and cabbage aphid, respectively. The eggs are usually elliptical in shape, initially yellow or green, but soon turn black. Nymphs look like to the viviparous adults. Winged aphids of *M. persicae* have a black head and thorax, and a yellowish green abdomen with a large dark patch dorsally. They measure 1.8-2.1 mm in length. The wingless forms are yellowish or greenish in color and about 1.7-2.0 mm long. They may have a medial and lateral green stripes. The cornicles are moderately long, unevenly swollen along their length, and match the body in color. The appendages are pale (Capinera, 2005). Wingless females of *B. brassicae* are about 2.1-2.6 mm long oval shaped and the posterior end of the body tapers greatly. The pale green body is usually covered by white waxy powder. Underneath the wax there are eight dark brown or black spots located on the upper abdominal surface. They increase in size toward the posterior end. Winged females are a little smaller than the wingless forms and they are not covered with waxy powder. The head and thorax are dark brown to black. The yellowish green abdomen is with two dark spots on the dorsal anterior abdominal segments. Antennae are dark brown. Wings are short and stout with prominent veins (Kessing & Mau, 1991).

Life cycle and damage: Eggs of *M. persicae* are deposited on *Prunus* spp. trees, where they overwinter. After several generations when the aphid densities increased the winged forms produced and disperse to the summer hosts. However, *B. brassicae* is restricted to herbaceous Brassicaceae throughout its life cycle. They lay their eggs on the stems and leaves of cruciferous crops that remain in the field through the winter. Both of these aphids overwinter in the egg stage. Generally aphids feed by sucking sap; infested seedlings may become stunted and distorted. Continued feeding on mature plants causes wilting, yellowing and general stunting of the plants. For peach potato aphid only very heavy infestations cause direct damage. Numbers may increase after serious outbreaks of insecticide-resistant aphids on potatoes. Peach potato aphids transmit the beet western yellow virus, which causes temporary reddening of infected plants in spring. In the case of cabbage aphid, spring rape can become severely infested after mild winters and allow infestations on winter rape to increase to damaging levels. The cabbage aphid is a vector of 23 virus diseases of Brassicaceae. It carries cauliflower mosaic virus. Also, turnip mosaic virus is carried by cabbage aphid and green peach aphid (Jellis, 2003).

Monitoring: monitor weekly from flowering to grain fill of the crop. At least four sampling points of random 20 plants should be spread over the field and estimate the number of aphids per plant.Also, yellow traps can use for monitoring aphids (Bailey, 2007).

Economic and action thresholds: *B. brassicae* action threshold is if over 13% of winter rape plants or over 4% of spring rape plants are infested before petal fall (Jellis, 2003). The threshold for *M. persicae* is correspondingly lower, at 25% of plants infested (Bailey, 2007).

Control: The parasitoid, *Diaeretiella rapae* (McIntosh) normally provides good control of cabbage aphid. Insecticide resistance has been observed for *M. persicae;* so expert advice on alternative insecticides should prepare. In the case of *B. brassicae* using a higher rate of pirimicarb plus added wetter is recommended (Jellis, 2003).

5.4 *Bagrada hilaris* (Burmeister) (Hemiptera: Pentatomidae)

Identification: painted bug adult is black with orange spots and markings called painted bug. These stink bugs are about 5-7 mm long. Females are larger than the males and both have distinctive black. The eggs are oval-shaped and off-white in color developing to an orange color when growth. Nymphs color changed as they develop from bright orange to red with dark markings, gradually acquiring the coloration of the adult. Wing pads are visible in the last instar nymph and gradually developed as they grow (Barlow, n.d.).

Life cycle and damage: Females lay about 50-80 eggs singly or in clusters of approximately 2-4 eggs during their lifetime. The eggs are laid on leaves or on the soil underneath host plants which hatch after 5-8 days. It has five nymphal stages. All stages of painted bugs can be found on a plant and both adults and nymphs suck sap from all parts of the plant. Young plants wilt and wither as a result of the attack. Also, adults excrete a resinous substance which spoils the pods. They often stunt the growth of newly formed central shoots or heads of plants. Quality and quantity (31% losses) of yield is affected when grown up plants are infected. Harvested crop in threshing floor is also infested (TIFP, n.d.-b)

Monitoring: monitor egg masses; if parasitized egg masses (black ones) are found, treatment for newly hatched nymphs might not be necessary. Also, monitor by beating or by shaking the vines on the cloth and count number of nymphs and adults.

Economic and action thresholds: In Africa where this insect comes from originally, the economic threshold is one bug/10 ft.^{2} on plants in the early growth stage increasing to 3 bug/10 ft.^{2} on mature plants (Barlow, n.d.).

Control: Deep ploughing; early sowing; irrigation of the crop and quick threshing of the harvested crop are the cultural techniques that can reduce pest attack. Burn the remains of mustard crop and conservation of bio-control agents like *Alophara* spp. can control painted bug population. Spraying malathion 50 EC, Endosulphan 35 EC, and Dimethothiate 30 EC should be considered if the threshold is exceeded (TIFP, n.d.-b).

5.5 *Psylliodes chrysocephala* (L.) (Coleoptera, Chrysomelidae)

Identification: Cabbage stem flea beetle is about 3-4.5 mm long, blue-black or light brown. Eggs are oval-shaped and about 0.9 x 0.4 mm, pale orange. The larvae are distinguished from other pests by their black head and black plate at the tail end (Jellis, 2003).

Life cycle and damage: females lay their eggs on the surface or in the cracks of the ground, singly or in clusters of 2 to 8 in the immediate vicinity of the stems of plants. The hatching larvae bore into petioles and continue feeding close under the surface. Later, larvae move into the main stem to feed under the growing point. The feeding injury appears as holes or small pits in the cotyledons and leaves. The beetles fly to canola, rapeseed and other mustards, moving into fields just as the seedlings emerge. They feed on the emerging cotyledon and first true leaves of the young plant. Feeding injury can result in plant death and stand loss. Flea beetles feed most actively when the weather is sunny, warm and dry. They overwinter as adult and have one generation per year.

Other flea beetles that are of increasing importance are: *Psylliodes luteola* (Müller) (Wessex flea beetle) which attacks crops in the autumn; *Phyllotreta cruciferae* (Goeze), *Phyllotreta nigripes* (Fabricius), (turnip flea beetles) attack crops in autumn and spring; and *Phyllotreta nemorum* (L.) (large striped flea beetle) which attacks crops in spring (Jellis, 2003).

Monitoring: mobile D-vac suction sampler can be used to sample cabbage stem flea beetles. Shoots enclosed within cylinders (0.25 m² area) are shaken and excised on the ground and subsequently the area should vacuum for three minutes. Water traps can be placed on the ground around field perimeters to monitor their immigration to the crop. Also, damage level should be determined for estimating time of management (Alford, 2003)

Economic and action thresholds: action threshold is over 25% of the leaf area at the 1–2 true leaf growth stage; or over 50% of the leaf area at the 3–4 true leaf stage; or when the crop is growing more slowly than it is being destroyed. Also, if over 5 larvae/plant are present in the field or over 50% petioles damaged (Jellis, 2003; Knodel & Olson, 2002).

Control: early planting can protect crops from pest's infestations. The larvae are biologically controlled by *Tersilochus microgaster* (Szépligeti). Seed Treatment with Chinook, Gaucho®, Helix®, Poncho® and Prosper® protect crops from early season infestation. If the threshold is exceeded application of a foliar insecticide reduce flea beetles population. Pyrethroid spray controls feeding adults and larvae that have not entered the main stem (Jellis, 2003).

5.6 *Meligethes aeneus* (F.) (Coleoptera: Nitidulidae)

Identification: Pollen beetles are about 2-3 mm long, and black with a hint of metallic green. The eggs are elongated, glassy appearance verging on the milky-white. The larvae are up to 3 mm long and have a head capsule and six legs (Jellis, 2003).

Life cycle and damage: females lay their eggs in the buds and after 4-9 days the eggs hatch. The common name of pollen beetle is because both adults and larvae feed on the flowers, pollen and nectar and contributes to the pollination of the crop. Loss of pod sites albeit sometimes severe, rarely reduces winter oilseed rape yields. After opening all buds of winter crops, beetles migrate to other egg-laying sites, causing a second flush of migration to spring crops. Crops usually compensate for earlier losses, producing more and larger seeds on lower racemes. The adult overwinters at the edge of woodlands or in undergrowth. *M. aeneus* has two generations per year. Varietal associations and restored hybrids may lose more yields because male fertile plants are attacked and crosspollination is reduced. Spring crops are much more vulnerable than winter crops (Jellis, 2003).

Monitoring: During flowering, the beetles are attracted by the yellow color of host plants. So, *M. aeneus* can be monitored by yellow traps. Also, beat the main flowering racemes of plants over a white tray and count insect's number (Alford, 2003; Ruther & Thiemann, 1997).

Economic and action thresholds: For winter oilseed rape the action threshold is when: over 15 beetles/plant at green-yellow bud stage; over 5/plant for backward crops; over 2/plant for varietal associations is recorded. In the case of spring oilseed rape observing over 3beetles/plant at green bud stage is the time of controlling pollen beetles (Jellis, 2003).

Control: *Phradis interstitialis* (Thomson), *Phradis morionellus* (Holmgren) and *Tersilochus heterocerus* (Thomson) can biologically control the larvae of pollen beetle. If thresholds are exceeded application of pyrethroid in green to yellow bud stages is advised (Jellis, 2003).

5.7 *Ceutorhynchus assimilis* (Paykull) (Coleoptera: Curculionidae)

Identification: Adults of cabbage seed weevil are matt grey about 2-3.5 mm long, with a distinctive, long, narrow, downward-curved rostrum (snout) on the front of the head. The prothorax has a notch in the middle of the underside front edge where the rostrum can rest. There are seven segments in the antennal funiculus. The elytra are black, but the elytral interstices have fine hairs and greyish white scales which results in an overall grey appearance. Near the mid-line of the elytra, the interstices have 1-3 irregular rows of scales along their length. There is no tooth on the hind femora and all tarsi are black to dark brown, similar in color to the femora and tibiae. The tarsal claws are simple, not toothed. The eggs are creamy white, smooth, cylindrical with rounded ends, and about 0.6 mm long by 0.4 mm wide. They are often covered with a mucus-like material. The larva is vermiform, white with a brown head capsule. Pupae are about 2 mm long and occur in earthen cells in the soil. They are initially white, but then turn yellow (CABI, 2008).

Life cycle and damage: Cabbage seed weevil is univoltine; adults emerge from overwintering sites in the spring and migrate to rape crops to feed and oviposit. Females lay eggs into developing oilseed rape pods during flowering. A brown scar indicates the egg-laying puncture, usually causing a kink in the pod. The larvae develop inside the pod and cause economic damage by eating the seeds. The abundance of cabbage seed weevil is greater in winter than spring. It has three larval instars (Jellis, 2003; Murchie et al., 1999).

Monitoring: installation of yellow water traps and beating the main flowering racemes of plants over a white tray and counting number of insects are techniques for monitoring cabbage seed weevil (Alford, 2003; Murchie et al., 1999).

Economic and action thresholds: action threshold is when over 0.5 beetle/plant in northern Britain or 1/plant elsewhere is observed (Jellis, 2003).

Control: Delaying sowing of spring crops reduces numbers of adults attracted and subsequently the number of laid eggs. *Trichomalus perfectus* (Walker), *Mesopolobus morys* (Walker) and *Stenomalina gracilis* (Walker) can biologically reduce the population. If the threshold is exceeded application of pyrethroid by petal fall and before too many eggs are laid can decrease the population buildup (Jellis, 2003).

5.8 *Ceutorhynchus quadridens* (Panzer) (Coleoptera: Curculionidae)

Identification: Cabbage stem weevil adult is about 2 to 3.5 mm long, ashy grey. The Elytra is dorsally with a white rectangular spot; striae much narrower than the interstices which are covered in whitish scales. Larvae are about 3 to 5 mm, relatively elongated, white; with small light yellow head (Evans, 2007).

Life cycle and damage: females lay eggs in the petioles, young stems or along the main veins of the leaves over a period of weeks, depending both on temperature and crop stage. The eggs hatch after 5-8 days. The larvae tunnel their way into the mid rib of the leaf or the stem. Some diseases such as stem canker can gain access to the stem through the holes made by larvae as they leave the plant to pupate. Larval growth lasts 3 to 6 weeks until pupation, which takes place in the ground. Adults cause no obvious damage but can be found on plants. They infest a large percentage of plants and reduce vigor and yield. They overwinter as adult under lumps of earth or under plant debris (Evans, 2007).

Monitoring: monitor directly the crop from emergence and look for signs of pests activity, presence of beetles, and shot holding of leaves.

Economic and action thresholds: ETL is when up to 26 larvae per plant is observed (Winfield, 1961).

Control: seed treatments by gamma-HCH or sprays of gamma-HCH, azinphos-methyl, azinphos-methyl plus demeton-S-methyl sulphone, chlorpyrifos or triazophos are effective against stem weevils (Graham & Gould, 1980). Carbofuran or phorate granules reduce larval infestations, but thiofanox granules and dimethoate sprays are ineffective (Evans, 2007).

5.9 *Delia radicum* (L.) (Diptera: Anthomyiidae)

Identification: Cabbage maggot or root maggot adults look like house flies but smaller, about 5 mm long, dark ash grey with a dark stripe along the top of the abdomen, and covered with black hairs and bristles (setae). The reddish purple eyes on males nearly touch in the centre of the head while female eyes are separated. In males, the presence of the basal brush of long setae (anteroventral setae) on the hind femur and by relatively shorter lateral setae of the 5th sternite processes. In the hind femur of females, the row of anteroventral setae is normally uneven and posteroventral setae are lacking. The eggs are white and about 1 mm diameter. The larva posterior extremity bears a crown of 10 small black membranous points, which is also visible on the pupae. The pupae are reddish brown (Wikipedia, 2011b).

Life cycle and damage: The adults feed on nectar and lay their eggs close to plants of the genus *Brassica* on cool, moist soil. The eggs hatch into white maggots after about six days and the larvae feed for about three weeks on the roots and stems of the cabbage plants. The presence of larvae cause delay in plant growth accompanied by the withering of leaves which develop a bluish tinge. Sometimes 300 larvae can be found on one plant, damaging the inner parts of the main root and disrupting the transport of water and nutrients to the stem and leaves causing the death of many plants. The larvae mature in about 3-4 weeks, then leave the roots and pupate about 5-20 cm deep in the soil. This species has one or two generation a year (Wikipedia, 2011b). Dosdall et al., (1994) stated that the susceptibility of *Cruciferae* species to *D. radicum* is different. Plants of *B. rapa* were most susceptible to *Delia* spp.; intermediate susceptibility was observed for plants of *B. napus* and *B. juncea*; and plants of *B. alba* were least susceptible. The mechanism of resistance by cruciferous species to infestation by *Delia* spp. is antixenosis or non-preference.

Monitoring: monitoring can be done by combination of yellow water pan traps and yellow sticky traps over an entire growing season (Broatch & Vernon, 1997).

Economic and action thresholds: 21 eggs per plant is economic threshold for two week old plants and about 100 eggs per plant, 4 weeks after planting (Bligaard et al., 1999).

Control: remove weeds from the field two weeks before planting. Plant when there are fewer flies around to deposit eggs. If infested plants are found, they should be removed to prevent contamination of other plants.

5.10 *Dasineura brassicae* (Diptera: Cecidomyiidae)

Identification: The adults are 1-1.5 mm long, female with reddish abdomen; male, blackish. The larva is about 1.5 mm long, white or yellowish-white. Pupa is inside whitish cocoon about 1.4 to 2.1 mm long, at a depth of 3 cm, in the soil (Pavela & Kazda, 2007).

Life cycle and damage: Brassica pod midge females lay eggs in clusters through holes left by seed weevils in developing pods. After 4 days the eggs hatch and many small, white larvae occur in each affected pod. Midges appear during flowering; they feed on the internal wall of the pod valve (or on the seeds) by discharging saliva. They cause swelling and eventually pod burst. The larvae have 3 instars and they develop through 9 to 15 days, when they drop to the ground, bury themselves and form cocoons. Some will pupate; others may undergo diapauses for 1 to 4 years. They have about 3 to 4 generation (Jellis, 2003).

Monitoring: monitor *D. brassicae* by yellow or suction traps. Cardboard delta traps with sticky base plates have been used to monitor males (Alford, 2003; Pavela & Kazda, 2007).

Economic and action thresholds: No thresholds for treatment exist (Jellis, 2003).

Control: Blocking oilseed rape fields and rotation around the farm will reduce the impact of immigrant pests. *Omphale clypealis* (Thomson) and *Platygaster subuliformis* (Kieffer) may attack pod midge, killing up to 75% of larvae (Jellis, 2003). Calypso 480 SC has effects as contact and feeding poison with systemic impact, a non systemic pyrethroid insecticide protect the plants from beginning of flowering to full flowering (Pavela & Kazda, 2007).

5.11 *Athalia proxima lugens* Klug (Hymenoptera: Tenthredinidae)

Identification: Adults are orange colored wasps with black head and smoky transparent wings. Larva is greenish black wrinkled body, with three legs of thoracic legs and seven pairs of prolegs. On slightest touch the larva falls to round and feigns death. The pupae formed in an earthen cocoon in the soil (Chowdhury, 2009; Navarajan, 2007).

Life cycle and damage: Mustard sawfly females lay eggs near the leaves margin. They attack the crop at early growth period when the seedlings are 3-4 week old. The larvae feed on the leaves of rapeseed and mustard making holes. In severe case larvae eat up entire lamina of leaf leaving behind the midribs. In extreme cases, this pest defoliates the plants completely. They devour the epidermis of the shoot, resulting in drying up of seedlings and failure to bear seeds in older plants. The yield loss is up to 5 to 18 % (Chowdhury, 2009).

Monitoring: monitor directly and count the number of larvae. Also, sweep netting and counting the number of adults can be used for monitoring sawflies.

Economic and action thresholds: no action threshold has been set for sawflies in mustard.

Control: Summer ploughing to destroy the pupae; early sowing; irrigation in seedling stage and severe cold reduce pest growth and are crucial for sawfly management. Conserving *Perilissus cingulator* (Morley) (parasites the grubs), and the bacterium *Serratia marcescens* (Bizio) infect the larvae of sawfly. Apply insecticides such as methyl parathion dust at the rate of 20-25 kg per hectare or Malathion 50 EC if the threshold is exceeded (TIFP, n.d.-b).

5.12 *Plutella xylostella* (L.) (Lepidoptera: Plutellidae)

Identification: Diamondback moth is small greyish with three white triangular spots along the inner-margin of the forewings. At rest the triangular markings of opposite wings appear as diamond shaped markings and hence the name. The larvae are pale green, body tapering slightly at both ends (TIFP, n.d.-b).

Life cycle and damage: females lay eggs on the lower leaves. The first instar larvae make tiny, irregular holes in the leaves. Diamondback moth is the most destructive pest of brassicasea throughout the world. Caterpillars feed on the foliage and make the leaves a withered appearance. Moths of later generations lay eggs higher on the plant and the larvae feed first on leaves, moving later to buds, flowers and developing seedpods. In later stages larvae bore holes in the leaves and the leaves may be eaten up completely. They also bores into pods and feeds developing seed (TIFP, n.d.-b).

Monitoring: monitor the presence of *P. xylostella* by pheromone traps or yellow sticky traps. Traps should be checked weekly and count the number of moths (TIFP, n.d.-b).

Economic and action thresholds: economic threshold is when 25-30 larvae observed per square foot or 1-2 larvae per plant, and there is significant evidence of damage to flowers or pods (Knodel et al., 2010).

Control: bio-control agents such as *Cotesia plutellae* (Kurdjumov), *Diadegma mollipla* (Holmgren), *Oomyzus sokolowski* (Kurdjumov), and *Apanteles* sp. can control diamondback moth. When the threshold exceeded application of malathion, Trichlorfon, Endoslfan 35 EC, Diazinon 20 EC or triazophos is recommended (Manyangarirwa et al., 2009; TIFP, n.d.-b).

5.13 *Pieris brassicae* (Linn) (Lepidoptera: Pieridae)

Identification: it is known as large white or cabbage butterfly. Forewings of the moth is mealy-white with dark pollination at base (wingspan 55-60 mm); females are larger than males. Wings at top with intensive black crescent limb reaching middle of outer margin and with 2 black rounded spots. They have sexual dimorphism and males just have 2 spots on the lower side of forewings. Hindwings have black stroke at anterior margin from above; grayish-yellow from below, with dense black pollination. The eggs are yellow skittle-shaped. The larvae are yellowish green with yellow lines and black spots. Pupae are greenish-yellow with black dorsal and lateral dots (Navarajan, 2007).

Life cycle and damage: females lay eggs in batches on the lower side of leaves. After 4 to 16 days the eggs develop. Young caterpillars aggregate and scraping the lower side of leaves; older ones live individually, gnawing holes in leaves, contaminating the latter with excrement. In severe attacks thick veins only leave. Caterpillars have 5 instars and develop in 13-38 days. The pupa phase lasts 8-15 days. Diapausing pupae do not hibernate far from host plant localities on tree trunks and branches, fences etc. They have the ability to migrate in large numbers during the spring and autumn and have 2-3 generation (Navarajan, 2007).

Monitoring: install pheromone traps for monitoring male moths and check the crop weekly after peak trap catches for the presence of larvae.

Economic and action thresholds: no action threshold has been set for *P. brassicae*.

Control: Caterpillars can be controlled biologically by *Apanteles glomeratus* (L.). Spray 0.05% dichlorvos, acetamiprid or 1% malathion if the threshold is exceeded (Navarajan, 2007).

5.14 *Hellula undalis* (F.) (Lepidoptera: Pyralididae)

Identification: Moths of *H. undalis* are pale greyish-brown, suffused with reddish color. Their forewings have wavy gray markings, a curved pale patch sub terminally, and a kidney shaped mark one third length from the tip. The wingspan is about 18 mm. Hindwings are pale, with the tip being lightly colored. The eggs are small, oval and slightly flattened upon the plant surface. They are creamy white when freshly laid, become pinkish the next day and then turn brownish-red. Larvae are pale whitish brown with 4 or 5 pinkish-brown longitudinal stripes. Pupation occurs in a silk cocoon and early stage pupae are soft and very pale yellowish-white in color with a bright red, dorsal blood (TIFP, n.d.-b).

Life cycle and damage: females lay eggs singly, or in groups or chains of 2 or 3 on the leaves near the bud and after 4 to 5 days the eggs hatch. Cabbage head borer caterpillars initially mine the leaves and make it white papery. Later they feed on leaves and bore into stems, entrance hole is covered with silk and excreta (TIFP, n.d.-b).

Monitoring: Regular monitoring of young plants in the nursery and after transplant is important. Install pheromone traps and detect larvae (Kessing & Mau, 2007).

Economic and action thresholds: ETL is when 15-25% of plants infested in a random pattern (Kessing & Mau, 2007).

Control: *B. hebetor* and *C. blackburni* can control *H. undalis*. Screening of seedling beds and clean culture are helpful in reducing damage caused by cabbage webworm. Biopesticides such as B.t. and spinosad are useful for the management of caterpillars. If the threshold is exceeded application of 5% malathion dust, Trichlorfon or carbaryl 50 WP, endosulfan (0.05%), cypermethrin (0.005%), and neem are advised (Navarajan, 2007; TIFP, n.d.-b).

5.15 *Crocidolomia binotalis* (Zeller) (Lepidoptera: Pyralidae)

Identification: leaf webber moths are yellowish-brown, forewings in parts have reddish-brown distinct and in distinct wavy lines and prominent white spots. Near to the thorax forewing and black tuft of hair and hindwings are white with dark brown apical area. The eggs are brown furry, flat and placed in masses. The larvae are pale yellowish-brown, with a series of lateral and sub-lateral black spots and specks. The pupa is yellowish green when formed and turns dark brown later and it is about 1 cm long (TIFP, n.d.-b).

Life cycle and damage: females lay eggs in masses and each mass containing 40-100 overlapping eggs. After 5-15 days, depending on whether conditions the eggs hatch. The newly hatched larvae feed initially on the chlorophyll of young leaves and later on older leaves, buds and pods, make webbings and live within. Severely attacked plants are defoliated and seeds in the pods are eaten away. Leaf webber is the secondary pest and may become a serious problem, particularly during the dry season. The larvae develop after about 24-27 days during summer; or about 51 days in winter. Leaf webbers pupate in an earthen cocoon, 2 to 6 cm below soil surface (TIFP, n.d.-b).

Monitoring: Monitoring is very important in the first stages of the crop. Check the crop weekly after peak trap catches in order to detect the caterpillars. Scout the plants with quarter-meter square (50 cm by 50 cm) and beat them within that area to dislodge the larvae. Count the fallen larvae and multiply by 4 to get the number per meter square (kalita, 2004).

Economic and action thresholds: ETL is one larva per meter row of length (kalita, 2004).

Control: field sanitation, crop rotation and intercropping are cultural techniques for controlling leaf webbers. *Aponteles* sp. is the pupal parasitoid of *C. binotalis*. Neem, B.t., carbaryl, cypermethrin, fenvalerate, or monocrotophos sprays are effective (kalita, 2004).

5.16 *Mamestra configurata* (Walker) (Lepidoptera: Noctuidae)

Identification: It is known as Bertha armyworms. The forewings of moths are predominantly gray, and flecked with patches of black, brown, olive and white scales. Near the middle of the forewing, there is a prominent, white, kidney-shaped marking defined with a ring of whitish scales. Near the tip of the forewing, there is a conspicuous white and olive-colored, irregular transverse marking that is characteristic of the species. The moth wingspan is about 4 cm. The eggs are sculptured, ridged; white at first but become darker as they develop. The larvae are pale green when they first hatch. Older larvae will be velvety brown to black with a yellowish band along each side of the body (Knodel et al., 2010).

Life cycle and damage: females lay eggs in single-layered clusters of about 50-500 eggs on the lower surface of the leaves. Firs instar larvae feed on the leaves. As leaves dried, the old larvae begin feeding on seeds and flowers which are more succulent. Early seeded canola often has been swathed prior to the occurrence of significant feeding injury. They cause serious damage by chewing through the stems below the bolls, causing them to drop to the ground. The larvae have six instars and at maturity, burrow into the ground and form pupae, their over-wintering stage. Bertha armyworm attacks many kinds of broadleaf plants, including canola, flax and beans (Knodel et al., 2010).

Monitoring: monitor the caterpillars and count the number of them on crops according to the method of leaf webbers.

Economic and action thresholds: Thresholds is when 18-22 larvae is observed per square yard, as long as leaf feeding is the extent of the damage observed. Thresholds may be adjusted lower if larvae are found feeding on maturing seed pods (Knodel et al., 2010).

Control: If the threshold exceeded apply insecticides such as: bifenthrin, deltamethrin, gamma cyhalothrin or lambda-cyhalothrin (Knodel et al., 2010).

6. Sunflower

Helianthus annuus L.: known as common sunflower, about 60% of the world's production is in Europe and the USSR (Union of Soviet Socialist Republics) (Putnam et al., 1990).

6.1 *Pseudoheteronyx basicollis* (L.) (Coleoptera: Scarabaeidae)

Identification: it is known as black scarab beetles or black sunflower scarab. Beetles are 13 mm long, shiny black with very short hairs. Eggs are cream and spherical, about 1.5 mm in diameter. The larvae are creamy with a grey rear end, brown head capsule and up to 25 mm long. They are C-shaped with wrinkly bodies and Y-shaped palidia (Franzmann, 2011).

Life cycle and damage: females lay their eggs under seedlings. Young larvae feed on taproots causing wilting and deaths of seedlings, older larvae attack the pods. Adult beetles can defoliate and kill plants up to 40 cm tall. Larvae overwinter in the soil and pupate in the spring. Adults often feed in a line across the field. Beetles hide in the soil during the day and emerge in late afternoon to feed. Heavily infested crops may suffer over 30% yield loss. They have one generation per year (Bailey, 2007).

Monitoring: check in the soil by digging and sieving for the presence of larvae prior to planting, and at all stages for adults. Count the number of beetles per m² (Brien, 2011).

Economic and action thresholds: action threshold is when four beetles is observed per square meter which can cause severe losses to young seedlings (Franzmann, 2011).

Control: Removal of the host parthenium weed is advised. Spray appropriate insecticides when beetles are active on the soil surface. Beetles can also be controlled by application of pelleted baits (alfalfa or similar meal) at planting. Damage is most prevalent where sunflowers follow wheat, sorghum or grass pasture (Brien, 2011).

6.2 *Dectes texanus* (LeConte) (Coleoptera: Cerambycidae)

Identification: Dectes stem borer is bluish-gray, about 1-2 cm, with antennae as long, or longer than the body. They are generally cryptic in behavior and can often be found hiding under leaves within the plant canopy during daylight hours, but also may be seen actively flying and running around on plant surfaces. The eggs are like small bananas with rounded ends. The larvae are pale, legless, cylindrical, and deeply segmented. They are creamy and tapered towards the rear so that final segments are narrower (Michaud et al., 2011).

Life cycle and damage: Females become reproductively active about one week after emergence, and require another week or so to mature eggs. Eggs are normally laid in leaf petioles deep in the central pith. After six to 10 days the eggs hatch and newly hatched larvae pass the first instar in the leaf petiole and then tunnel down into the main stalk where they feed selectively on the pith in the central core. Larvae develop through six instars. The effect of larval feeding on yield is negligible in healthy plants, but reduces plant resistance to other insects such as stem weevils. As stalks dry down, mature larvae of *Dectes* descend to the base of the stem, begin to girdle the interior surface near the soil line, and then plug the tunnel with chewed fibers that resemble sawdust before retreating to the base of the plant to overwinter. Plants weakened by complete or partial girdling snap off easily when pressure is applied laterally to the stem. Adults mate and feed on plants, leaving longitudinal feeding scars on stems and petioles. Infestations approaching 100 percent of plants may go completely unnoticed when fields are harvested early, but any delay in harvest can result in serious losses due to lodging. They have only one generation per year (Michaud et al., 2011).

Monitoring: monitor in late summer by splitting stalks to determine if Dectes stem borers are present, and harvest as soon as possible if infestation is extensive (Michaud et al., 2011).

Economic and action thresholds: no action threshold has been set for Dectes stem borer.

Control: Crop rotation may reduce damage from Dectes when the acreage of soybeans and sunflowers in an area is low. Stalk desiccation is an important cue triggering the girdling behavior of *D. texanus* larvae. Later planting dates, fall or winter tillage and irrigation of sunflowers have reduced infestations by this pest. There is no varietal resistance to stem borers in sunflowers and chemical treatments are not recommended. Although adults are susceptible to many foliar insecticides, their extended activity period means that a single application will not provide control. However, larvae feed within the plant and are protected from contact insecticides, and even repeated applications of systemic materials fail to provide adequate control in most years (Knodel et al., n.d.; Michaud et al., 2011).

6.3 *Cylindrocopturus adspersus* (LeConte) (Coleoptera: Curculionidae)

Identification: Sunflower stem weevil adults are less than 4-5 mm long and grayish brown with white markings on the wing covers and thorax. The snout, eyes and antennae are black. Eggs are very small about 0.51 mm long by 0.33 mm wide, oval and yellow, and are difficult to see. Larvae are approximately 6 mm long at maturity. Larvae are legless and creamy white with a small, brown head capsule. They are normally in a curled or C-shaped. Pupae are similar in size to the adult and are creamy white (Knodel et al., n.d.).

Life cycle and damage: adult emerge from overwintered stalks or root crowns and mating occurs soon after emergence. Adults feed on stem and leaf tissues which causes minor damage to them. Females lay eggs individually in the stem tissue around the first node (cotyledon) or inside the epidermis of sunflower stems. First instar larvae feed on sub epidermal and vascular tissue in the stem. Feeding is concentrated in the pith tissue as the larvae develop to third and fourth instar stages. It has about five to seven larval instars. Larvae feed apically and last instars descend to the lower portion of the stalk or root crown and excavate overwintering chambers by chewing cavities into the stem cortex. They have only one generation per year. The stem may break, causing a loss of the entire head prior to harvest. Stalk lodging is most severe during drought stress or when high winds occur as plants are drying prior to harvest. Lodging typically occurs at or slightly above the soil line, in contrast to breakage attributed to stalk diseases. Lodging is a good indicator of larval densities; however, lodging is influenced by other factors, including stalk diameter, cortex and pith thickness of the stem, weight of sunflower heads, wind velocity and direction, position of larvae in overwintering chambers in the stalk (Knodel et al., n.d.).

Monitoring: monitor the field and count number of weevils. Adults can be found on both surfaces of the leaves, the lower portions of the stem, in leaf axils, within the dried cotyledons or in soil cracks at the base of the plant. Field scouting for adults should begin when plants are in the eight to 10 leaf stages. Five randomly sampling sites are determined 70-100 feet in from the field margin and in each sample site, five plants is selected. It is recommended to use an X pattern (or W pattern) to space sample sites throughout the entire field (Knodel et al., n.d.). According to the Michaud et al., (2011) scouting adults is really difficult and not recommended in many cases. So, in a region with a history of stem weevil problems growers may assume that any dry land sunflowers will be at risk, and that their impact on the crop will increase with adversity of growing conditions.

Economic and action thresholds: The economic threshold is one adult per three plants, or about 40 larvae per stalk at the end of the season (Knodel et al., n.d.).

Control: delayed planting and reducing plant population increases stalk diameter which is useful for controlling *C. adspersus* (Knodel et al., n.d.). Systemic pesticides such as carbofuran give simultaneous control of stem weevils and most other stem-infesting insects, and may also preserve stalk integrity by delaying processes of deterioration. Although seed treatments can be useful for improving seedling establishment, their duration of activity within the plant is not sufficient to provide control of stem weevils. Applying insecticides such as Beta-cyfluthrin, Chlorpyrifos, Chlorpyrifos plus gamma-cyhalothrin (Cobalt), Chlorpyrifos plus zeta-cypermethrin, Deltamethrin, Esfenvalerate, and Gamma-cyhalothrin (Proaxis) is recommended if the threshold is exceeded (Michaud et al., 2011).

6.4 *Apion occidentale* (Fall) (Coleoptera: Curculionidae)

Identification: It is known as black sunflower stem weevil. Adults are shiny black and less than 3 mm long. The snout is very narrow and protrudes forward from the head, which is small in relation to the rather large, almost globosely body. Larvae are similar in appearance to the sunflower stem weevil larvae, except they are only 3 mm long at maturity and more yellow with a more pointed posterior (Knodel et al., n.d.).

Life cycle and damage: Adults overwinter in soil or plant residue; after they emerge they feed on leaf and stem tissue. Females lay eggs under the epidermis in leaf petioles or stems near leaf axils. Larvae feed in vascular and pith tissues of stems and petioles. Newly emerged larvae tunnel in the pith of the stem, pupate and emerge as adults. second generation adults feed on the leaves and stems of the plant, but as the plant matures and the leaves begin to die, adults move under the bracts of the sunflower head, where they can be observed feeding until the plants are harvested. Adults move into the soil to overwinter. Although feeding damage is seldom significant, however, this species has been associated with the transmission of the pathogen *Phoma macdonaldii* (Boerma), the causal agent of phoma black stem. Stand loss can occur where extremely high populations are feeding on sunflower seedlings. In most cases, however, populations are too low to cause economic damage, and stalk tunneling only results in minor injury to the plant (Knodel et al., n.d.).

Monitoring: refer to the monitoring section for *C. adspersus*.

Economic and action thresholds: no action threshold has been set for this species.

Control: This species, has not been considered as an economically important pest in cultivated sunflower fields. Controlling other pests such as *C. adspersus* appears to give simultaneous control of *A. occidentale*.

6.5 *Baris strenua* (LeConte) (Coleoptera: Curculionidae)

Identification: Sunflower root weevil adults are rather robust weevils, with a somewhat oval-shaped body. Adults are dull black and 6 mm long with a short, almost blunt, downward projecting snout. Larvae are 6 mm long at maturity, legless, and have a white body with a small, brown head capsule (Knodel et al., n.d.).

Life cycle and damage: Adults feed on sunflower foliage in early morning and late afternoon. About two weeks after emergence, adults begin to congregate around the root zone near the soil surface. They continue feeding and mating occur during this period. Feeding activity produces callus tissue, under which the bright yellow eggs are deposited two or three at a time. First instar larvae are not very mobile and feed on the epidermal and cortical cells of the roots. Most feeding (consisting of circular tunnels) and development to fourth instar takes place in the same area where egg hatching occurs. In the fourth larval stage, the plant becomes significantly dehydrated and encapsulation of the larvae within a soil cocoon begins. Larvae overwinter within the cocoon among the remaining roots in the soil and it has one generation per year (Knodel et al., n.d.). When larvae are exceptionally abundant, they may cause plant wilting and lodging (Michaud et al., 2011).

Monitoring: monitoring is not so important for this pest species.

Economic and action thresholds: No economic thresholds have been established for root weevils.

Control: There are currently no management recommendations for sunflower root weevil. However, growers should realize that significant burrowing in sunflower fields and numerous uprooted plants are often caused by mammals foraging for root weevil larvae (Michaud et al., 2011).

6.6 *Smicronyx sordidus* (LeConte) and *Smicronyx fulvus* (LeConte) (Coleoptera: Curculionidae)

Identification: They are known as the gray seed weevil, and the red seed weevil, respectively. The gray seed weevil is about 8 mm long and pale gray, with a gently curving snout. Adults can frequently be found hiding under the bracts of sunflower buds either prior to or during bloom. *S. fulvus* is a small, reddish-brown weevil, about 3-4 mm long. Adults are most easily seen on the faces of blooming sunflowers feeding on pollen, or hiding behind the flower bracts. Infested seeds are not noticeably different from healthy ones, but display an exit hole near the top once the larva has matured and left (Michaud et al., 2011).

Life cycle and damage: *S. sordidus* females favor flowers in the late bud stage for oviposition, laying their eggs singly at the base of the developing seeds. Upon hatching, larvae bore into the seeds at their base, destroying the ovaries. At maturity, larvae exit the seeds and drop to the ground where they seek overwintering sites in the soil. The larvae feed within developing seeds and cause a galling response, leaving affected seeds swollen in appearance, but empty. Adult weevils of *S. fulvus* seek sunflowers in bloom and females must feed on pollen for at least four days before they can produce eggs. Females lay egg individually and directly into a seed and this requires a seed in the appropriate developmental stage. Developing larvae do not usually consume the seed completely, but significantly reduce kernel weight and oil content. The time frame of larval development is such that a percentage of larvae may still be present in seed at harvest, leading to temperature and moisture problems in storage. Larvae often can be found in seeds at harvest, especially when harvest is early, but most die in storage without causing further damage. They can result in yield reductions. However, the red seed weevil is more commonly responsible for economic damage than the gray seed weevil. Both of them have one generation per year (Michaud et al., 2011).

Monitoring: monitoring should be done as the field approaches 85 % bloom, or when 80 % of plants are past the R4 stage. Count the number of weevils on the faces of five of the most mature flowers at each of five sites. This process can be facilitated by spraying the faces of each flower with a DEET-containing mosquito repellent that will cause weevils to quickly exit the flower. For more information refer to the monitoring section of *C. assimilis* in mustard (Michaud et al., 2011).

Economic and action thresholds: No economic threshold has been established for the gray seed weevil and it is generally considered a less serious pest than the red seed weevil. Thresholds calculated for red seed weevil are sometimes used for gray seed weevil and control of gray seed weevil requires pesticide applications prior to bloom, when about 10-15% of plants have reached the R4 stage. Generally, the threshold for oilseed sunflowers falls somewhere between 10 and 20 weevils per flower, but the threshold is much lower for confection sunflowers (usually 1 or 2 weevils per flower) because of industry standards that demand seed damage remain below 3-4 % of kernels. Research suggests that approximately 27 damaged kernels result for every adult weevil observed at the early bloom stage. An early treatment should be followed by continued scouting at 4-5 day intervals, as adult emergence continues over several weeks (Michaud et al., 2011).

Control: tillage has been shown to reduce survival of overwintering weevils. Chemical treatments are recommended if the threshold is exceeded. Beta-cyfluthrin (Baythroid XL); Chlorpyrifos; Chlorpyrifos plus gamma-cyhalothrin (Cobalt); Chlorpyrifos plus zeta-cypermethrin (Stallion); Deltamethrin (Delta Gold); Gamma-cyhalothrin (Proaxis); Lambda-cyhalothrin, Parathion, methyl (Cheminova Methyl 4EC) are appropriate for controlling seed weevils (Michaud et al., 2011).

6.7 *Pterohelaeus alternatus* (Pascoe), *Pterohelaeus darlingensis* (Carter) and *Gonocephalum macleayi* (Blackburn) (Coleoptera: Tenebrionidae)

Identification: They are known as striate false wireworm, Eastern false wireworm and Southern false wireworm, respectively. Adult beetles of *Pterohelaeus* spp. are 20 mm long and dark grey-black with a distinctive 'pie-dish' shape formed by flanges around the outline of the beetle. Adults of *Gonocephalum* spp. are 9 mm long, dark grey-black and often covered in soil. There are flanges around the outline of the thorax (behind the head). Larvae are up to 30 mm long, shiny and cream, yellow or tan with three pairs of legs behind the head. They are hard-bodied, cylindrical and segmented with a rounded head (Murray, 2010).

Life cycle and damage: females lay eggs singly in moist soil, usually under trash or low-growing weeds. Both adults and larvae attack sunflower and damage of them may necessitate replanting. The larvae feed on decaying vegetable and crop residues in the soil; they also feed on newly germinating seed and the growing points of seedlings which results in patchy stands. Damage is most common in early-planted crops where crop residue has become scarce. They are major pests of seedling field crops and especially attack sunflower, soybeans, and mungbeans. They usually have one generation per year (Brien, 2011).

Monitoring: monitoring can be difficult; either hand sift 10 soil samples (30 x 30 cm) or place 10 germinating seed baits monitoring for soil-dwelling insects throughout the paddock. Pitfall traps can be used for counting the number of false wireworms (Brien, 2011).

Economic and action thresholds: The action threshold is observing more than 25 wireworm larvae in 20 germinating seed baits (Murray, 2010).

Control: Use of press wheels at planting; clean cultivation during summer can be effective against wireworms. Control of adults is obtained by baiting with insecticide-treated cracked and larvae can be controlled by insecticide applications at planting or insecticide-treated seed (Brien, 2011).

6.8 *Strauzia longipennis* (Wiedemann) (Diptera: Tephritidae)

Identification: Sunflower maggot is the only tephritid species found in the stalks of cultivated sunflowers. The showy yellow adult has a wing span of about 13 mm and a body length of 6 mm. The eyes are bright green and the wings bear broad dark bands that form a fairly distinct F pattern near the wing tip. Eggs are 1 mm long and are white and elongated. Larvae are yellow-white, headless and legless, tapered from anterior to posterior and approximately (7 mm) long at maturity (Knodel et al., n.d.).

Life cycle and damage: Adults emerge and flies have been noted in shelterbelts and field margins. Females lay eggs singly in stem tissue near the apical meristem, and larvae feed in the stalk pith, creating large tunnels. The maggots, when fully developed, emerge from the stalk. Larvae develop through three instars in approximately six weeks. This species overwinters as a larva in plant debris in the soil. The sunflower maggot has one generation per year. Feeding is confined to the pith, which acts as a supporting structure, and is not critical to plant nutrition. Secondary fungal infections also are associated with tunneling by the larvae within the stalk. Stalks are not weakened and seed yield is not reduced, even with severe pith destruction (Knodel et al., n.d.).

Monitoring: no scouting methods have been established for sunflower maggot.

Economic and action thresholds: no thresholds has been established for sunflower maggot.

Control: controlling other pests of sunflower will manage maggot infestations.

6.9 *Trichoplusia ni* (Hübner), *Thysanoplusia orichalcea* (F.) (Lepidoptera: Noctuidae)

Identification: *T. ni* is known as cabbage loopers. Moths are dark, smoky, and gray variegated with light grayish brown. Characteristic small silvery oval spots and U-shaped silvery white marks are on the middle of the forewings. Hind wings are pale brown with the veins conspicuously visible. Males have tufts of gold hair at the tip of the abdomen. Eggs are hemispherical, pale green or white with shallow ridges that meet at the center of the egg. The newly emerged caterpillars are translucent white. Once feeding starts the larvae become pale green, with a thin white line running lengthwise down each side of the body and two white lines along the middle of the back. Pupae are yellow green with a few brown patches when newly formed and gradually darken to dark brown before adult emergence. The pupation occurs in white transparent silken cocoon in folded leaf (Mau et al., 2007). *T. orichalcea* is known as soybean looper, but larvae are an occasional pest of sunflower. The moth forewings are brown with a large, bright gold patch. Hind wings are fawn-colored, darkening towards the outer margins. The Eggs are pale yellow-green, ribbed and flat. Very small larvae are green all over but medium larvae usually have prominent dark and white striping. Larvae pupate under leaves in a thin silken cocoon. Pupae are dark above and pale underneath (Brier, 2010).

Life cycle and damage: females of *T. ni* lay eggs singly on the lower leaf surface near the leaf margin. Eggs hatch in 3-4 days. This is one of the major pests of sunflower. The young larvae feed on chlorophyll contents and make part of leaf transparent. As the larvae grown they feed on the leaf margins and during severe infestation only midribs are left. The presence of small transparent leaf spot, devoid of chlorophyll symbolizes the beginning of pest population. The caterpillar stage lasts for 12-20 days. The pupal stage lasts for 9 days

(Mau et al., 2007). *T. orichalcea* small larvae feed on only one side of the leaf. Developed larvae chew holes in the leaf, and then feed from the leaf margin. Larvae are primarily foliage feeders in soybeans but will attack the flowers and developing pods (Brier, 2010).

Monitoring: Beat-sheet is preferred sampling method for loopers. Also, light and pheromone traps can be used for monitoring and determining the moths peak (Brier, 2010).

Economic and action thresholds: An action threshold for *T. ni* is observing 0.5 cabbage looper per plant. In the case of *T. orichalcea* in pre-flowering crops, looper control is warranted if defoliation exceeds 33%. Tolerable defoliation drops to 15-20% once flowering and podding commences (Brier, 2010; Maxwell & Fadamiro, 2006).

Control: loopers can be controlled biologically by releasing *Copidosoma floridanum* (Ashmead), *Voria ruralis* (Fallen), *Hyposoter exiguae* (Viereck), *Enicospilus* sp. and *Disophrys lutea* (Brulle). Also, conservation bio-control agents like green muscardine fungus, NPV, B.t. may be effective for reducing population build up. If the threshold is exceeded applying insecticides such as quinalphos 25 EC and Endosulfan 4% is recommended (TIFP, n.d.-c).

6.10 *Vanessa cardui* (L.) (Lepidoptera: Nymphalidae)

Identification: The painted lady adult has a pointed forewing with a distinctive transverse white bar. The intensity of reddish coloration on the wings varies greatly, with some individuals brightly colored and others appearing quite drab. The eggs are green and barrel-shaped. The larvae are spiny, covered with bristles, and range in background coloration from mottled, pale green to dark, purplish hues (Michaud et al., 2011).

Life cycle and damage: During summer, loosely-knit swarms migrate northward in search of suitable food plants. When large numbers of adults arrive in sunflower fields there is the potential for substantial defoliation to occur. Females lay eggs singly on the upper surfaces of leaves and the eggs hatch after 3-5 days. The larvae feed singly, skeletonizing leaves and tying up silken nests that typically accumulate piles of larval frass. So, damage is normally confined to individual plants, but when many plants are affected, an effort to estimate percent defoliation in the field may be justified. The caterpillar takes 7-11 days to turn into a chrysalis and 7–11 days for the chrysalis to turn into a butterfly (Michaud et al., 2011).

Monitoring: monitor by walking and examining 20 plants at each of five separate locations and determine percentage of defoliation in each location (Michaud et al., 2011).

Economic and action thresholds: action threshold is when plants have sustained 25% defoliation and larvae are still less than 3 cm long (Michaud et al., 2011).

Control: Applying insecticides is advised if the threshold is exceeded. Chlorpyrifos plus gamma-cyhalothrin (Cobalt); Chlorpyrifos plus zeta-cypermethrin (Stallion); Gamma-cyhalothrin (Proaxis) and Lambda-cyhalothrin are recommended (Michaud et al., 2011).

6.11 *Chlosyne (Charidryas) nycteis* (Doubleday) (Lepidoptera: Nymphalidae)

Identification: Silvery Checkerspot is similar in appearance and biology to the painted lady, just the silvery checkerspot has more yellow and less red coloration on the wings. The lower surface of the hind wing bears a large, white crescent on its margin. The larvae are dark and covered with branching bristles with a distinctive yellow line running down each side of the body. There are initially gregarious, but disperse once their initial food plant is consumed (Michaud et al., 2011).

Life cycle and damage: Females lay their eggs in large clusters, often as many as 100. Even large plants can be completely skeletonized by larval feeding, but more damage can occur when egg clusters are laid on young plants in the V4-V6 stage, in which case the entire plant is quickly consumed and larvae migrate to neighboring plants, sometimes creating a bare hole in the field. They have typically two generations per year (Michaud et al., 2011).

Monitoring: refer to the *V. cardui* monitoring in sunflower.

Economic and action thresholds: The action threshold for defoliation of established plants is 25%, but treatment can only be justified if larvae are still less than 3 cm long, as larger larvae will soon cease feeding (Michaud et al., 2011).

Control: Use materials registered for painted lady caterpillars on sunflower.

6.12 *Euxoa messoria* (Harris), *Euxoa ochrogaster* (Guenee), *Feltia jaculifera* (Walker) (Lepidoptera: Noctuidae)

Identification: They are known as darksided cutworm, redbacked cutworm and dingy cutworm, respectively. The adult of darksided cutworm has light and grayish brown forewings with indistinct markings. Larvae are pale brown dorsally and white on the ventral areas. The sides have numerous indistinct stripes. At maturity, they are about 32 to 38 mm long and 6 mm wide. Redbacked cutworm moth has reddish brown forewings with characteristic bean-shaped markings. Larvae are dull gray to brown and are about 25 to 32 mm long when mature. Larvae can be distinguished from other cutworm species by two dull reddish stripes along the back. Dingy cutworm forewings are dark brown with bean-shaped markings as in redbacked cutworm adults. Hind wings of the male are whitish with a broad, dark border on the outer margin; in the female, they are uniform dark gray. Larvae are dull, dingy brown body mottled with cream color and have a thin light line down the middle of the back with a series of diagonal markings on either side (Knodel et al., n.d.).

Life cycle and damage: Females of darksided and redbacked cutworm moths lay their eggs in the soil. The larvae continue to feed and grow until when mature; mature larvae pupate in earthen cells near the soil surface. The pupal period lasts about three weeks. Both species have one generation per year. Adult dingy cutworms lay eggs on plants in the family Asteraceae in the fall. Larvae develop to the second or third instar in the fall and overwinter in the soil. It has one generation per year. Due to the larval feedings seedlings being cut off from 25 mm below the soil surface to as much as 25-50 mm above the soil surface. Young leaves may be severely chewed from cutworms (notably the darksided cutworm) climbing up to feed on the plant foliage. During the daytime, cutworms usually are found just beneath the soil surface near the base of recently damaged plants and feed at night. Wilted or dead plants indicate the presence of cutworms. Cut plants may dry and blow away, leaving bare patches in the field as evidence of cutworm infestations (Knodel et al., n.d.).

Monitoring: monitoring should be done as soon as sunflower plants emerge, and fields should be checked at least twice per week. A trowel or similar tool should be used to dig around damaged plants to determine if cutworms are present. The Z pattern should be used to determine cutworm population levels by examining five 1-square-foot (30 by 30 cm) soil samples per site (in the row) for a total of 25 samples (Knodel et al., n.d.).

Economic and action thresholds: The economic threshold is one larva per square foot (30 by 30 cm) or 25-30% stand reduction (Knodel et al., n.d.).

Control: Several insecticides are registered for cutworm control in sunflowers. Post emergent treatment with an insecticide provides quick control of surface feeding cutworms. Best results occur if insecticide applied at night. Seed treatment will provide suppression of cutworm activity (Knodel et al., n.d.).

6.13 *Psittacula krameri krameri* (Scopoli), *Psittacula krameri manillensis* (Bechstein) and *Psittacula krameri borealis* (Neumann)

Identification: They are known as African ring-necked parakeet, Indian ring-necked parakeet and Neumann's ring-necked parakeet, respectively. *P. K. krameri* is green, face, abdomen and under wing-coverts yellowish-green; nape and back of head variably washed with blue; chin, broad cheek-stripe; narrow band to nape pink; upperside of middle tail-feathers blue with greenish-yellow tips, outer feathers green; underside of outer tail-feathers olive-yellowish, middle feathers blackish; bill blackish-red with black tips; iris yellowish-white; feet greenish-grey. *P. K. manillensis* is much darker; face stronger green; blue tinge to nape extends in many birds to back of head; breast and abdomen feathers tinged bluish-grey; upper mandible red, lower mandible black; larger. However, *P. K. borealis* is as *krameri*, but face pale green; breast and abdomen feathers in both sexes with marked grey-white tinge; smaller upper mandible red with black tip (Rana, n.d.).

Life cycle and damage: They mostly observed in small groups. They clutch 2-6 eggs; which incubated in 21 to 24 days; and fledging period is about 6 to 7 weeks. The birds' damage starts from the milky stage and continues till harvest. They consume an average of 152 seeds per day. The damage of the birds was highest at the mature stage, 23.8%, and minimum, 7.60%, on the emerging stage of sunflower. Also, other birds such as sparrows, crows and blackbirds can damage sunflower crops (Ahmad et al., 2011; TNAU, n.d.-b).

Control: Establishment of scare crows in the field; bursting of crackers and carbide guns, tying polythene bags may decrease birds' infestations. Destruction of bird nest in and around the field and deploying two laborers per hectare to scare away the birds may be effective. Spraying of neem kernel powder solution at 10 g/liter of water after seed shedding for repelling the birds can be applied (TNAU, n.d.-b).

7. Safflower

Carthamus tinctorius L.: known as safflower, Safflor, Bastard saffron. Safflower has worldwide production, especially in several Western states and Canadian Prairie provinces.

7.1 *Camnula pellucida* (Scudder), *Melanoplus packardii* (Scudder), and *Melanoplus sanguinipes* (F.) (Orthoptera: Acrididae)

Identification: The clear-winged grasshopper, *C. pellucida* is a small but severe species of grasshoppers. *M. packardii* are strong fliers with long wings; the striped sand grasshopper. *M. sanguinipes* is named the Nevada sage grasshopper (Zinni, 2011)

Life cycle and damage: The *Melanoplus* species fed preferentially on leaves, floral parts, and capitula, while *C. pellucida* exhibited only peduncle feeding, which resulted in head clipping.

Defoliation of 20 to 30% was associated with significant increases in total dry matter, seed yield, and number of capitula. Further defoliation resulted in decreases of dry weight, seed yield, and capitula number (Mundel & Johnson, 1987).

Monitoring: refer to the monitoring section for grasshoppers in cotton.

Economic and action thresholds: action threshold is when 8 or more grasshoppers per square yard occur in the field (Knodel et al., 2010).

Control: refer to the control section for grasshoppers in cotton.

7.2 *Dactynotus carthami* (Hille Ris Lambers) [basionym] (Homoptera: Aphididae)

Identification: the synonym of safflower aphids is *Uroleucon compositae* (Theobald). Adults are about 1.5-2 mm long, black with pear-shaped body and conspicuous cornicles. Nymphs are reddish brown (Martínez, 1999).

Life cycle and damage: It is a serious pest of safflower. During pre-flowering stage both nymphs and adults suck the cell sap from shoot apices, peduncles, leaves and stem, secrete a honey dew like secretion on upper surface of the leaves and plant parts forming a black sooty mold which hinders photosynthetic activity resulting in stunted growth. In sever attacks the plants dry up. Yield losses cause by aphids is about 40-50% and infestation may occur 30-45 days old crop. It has four nymphal stages (TNAU, n.d.-a). There are other major aphids species that cause severe or moderate damage to safflower such as *Uroleucon compositae* (Theobold), *Dactynotus orientalis* sp., *Dactyonotus jaceae* (Linn.), *Macrosiphum sonchi* (H.R.L.), *Macrosiphum sonchi* (Linn.), *Macrosiphum compositae* (Theobold), *Macrosiphum* spp.(jaceae), *M. persicae*, *Aphis fafia* (Scop) and *Capitophorus eleagni* (Del Guercio) (Hanumantharaya et al., 2008).

Monitoring: install yellow traps for monitoring aphids.

Economic and action thresholds: The economic threshold level is observing 50-60 aphids/5cm twig/plant (Martínez, 1999). However, the economic threshold level differs according to the variety. For *U. compositae* ETL was determined 48.78 aphids/5 cm apical twig/plant on Bhima variety of safflower, with exposure periods of about 2-3 weeks from first aphid incidence. On Annigeri-1 variety of safflower, the ETL was estimated at 38.5 aphids/5 cm apical twig/plant with exposure periods of 2-3 weeks from first aphid appearance (Hanumantharaya et al., 2008).

Control: planting resistant varieties and early sowing escape the peak pest incidence. Application of balanced fertilizer, intercropping and mixed cropping reduces the pest population. Intercultural operations like harrowing; hoeing can reduces the weeds which are host plants for safflower aphids. Also, releasing natural enemies can be used in augmentation programs (Hanumantharaya et al., 2008).

7.3 *Acanthiophilus helianthi* (Rossi) (Diptera: Tephritidae)

Identification: Capsule fly or safflower bud fly adults are ash colored; black obscured by a dense gray microtrichia and wing with characteristic diffused pattern. They have light brown legs. Maggot is dirty white in color (Martínez, 1999).

Life cycle and damage: females lay eggs in clusters of 6 to 24 inside the young flower buds which hatch in 1-3 days. Newly hatched larvae feed on the soft parts of the capsules and later instars feed on the soft part within. Larval period lasted about 7 days in early summer. The infested buds rot and give an offensive smelling fluid. Losses leads to disrupted plant activities, reduction in flower buds, and, ultimately, to decreased quality and quantity of crop. The flies spent their entire lifespan from egg to adult inside the flower heads of safflower plants. Pupation takes place in flower buds and lasted for 5-7 days. The fly overwintered as pupae inside flower heads left in the field after harvest. They have three generations per year. The pest infestation leads to a yield loss ranging from 38.6 to 93.2% (Saeidi & Nur Azura, 2011; TNAU, n.d.-a).

Monitoring: modified Steiner traps baited with methyl eugenol, water traps and sticky traps can be used to monitor bud flies and their immigration to the crop (Alford, 2003).

Economic and action thresholds: no action threshold has been set for safflower bud fly.

Control: sowing of the crop; clean cultivation and using resistant varieties are cultural techniques for controlling safflower bud flies. Some bio-control agents such as *chrysopa* sp., *Orymurus* sp., *Eurytoma* sp., *Stenomalus muscarum* (L.), *Syntomopus* sp., *Bracon* sp., *Pronotalia* sp. and *Antistrophoplex conthurnatus* (Masi) can reduce the population build up. Application of insecticides such as dimethoate 30 EC, malathion 50 EC and phosphomidon 100 EC is recommended if the threshold is exceeded (TNAU, n.d.-a).

7.4 *Heliothis armigera* (Hübner) (Lepidoptera: Noctuidea)

Identification: *H. armigera* is known as gram pod borer or capsule borer. Refer to the identification section for *H. armigera* in cotton.

Life cycle and damage: The life cycle is more or less the same as on cotton but differ due to the weather conditions. In early stage of crop growth larvae feed on leaves and shoot apices. Later, the larvae shift to the developing capitula. The symptoms are perforated leaves, perforated involucres bracts, partially or completely eaten capitula in the bud stage and bored developing capitula (TNAU, n.d.-a).

Monitoring: refer to the monitoring section for *H. armigera* in cotton.

Economic and action thresholds: The ETL of capsule borer is 50 larvae per 100 plant (TNAU, n.d.-a).

Control: refer to the control section for *H. armigera* in cotton.

7.5 *Perigea capensis* (Guen) (Lepidoptera: Noctuidea)

Identification: the synonyms of the species are *Perigea illecta* (Walker) and *Platysenta illecta* (Walker). The adult is dark brown and forewings are dark brown with pale wavy marks; Hind wings are light brown. Safflower caterpillars are stout, green and smooth. The anal segment is humped and the body has some purple markings (TNAU, n.d.-a).

Life cycle and damage: The larva feeds on the leaves, bracts, flowers, capsules and sometimes on capitulum too. They also bore into the stem and flower-buds and feed on the contents (TNAU, n.d.-a).

Monitoring: larvae can be monitored by directly observations and counting the number of larvae per plant on 10 randomly selected and tagged plants.

Economic and action thresholds: no action threshold has been established for *P. capensis*.

Control: Intercropping with non-host crop like wheat can be effective. They can be controlled biologically by *Apanteles ruficrus* (Haliday), *Hetergamus* sp., *rogas* sp. and *Euplectrus euplexiae* (Rohwer), green muscardine fungus (*Metarhizium anisopliae*) and NPV. If the threshold is exceeded spraying Indoxacarb 15 EC, Spinosad 45 SC, Carbaryl 50 WP, Endosulfan or Fenvalerate 20 EC is advised (TNAU, n.d.-a).

7.6 *Spodoptera exigua* (Hubner) (Lepidoptera: Noctuidae)

Identification: moths have a wingspan of 25-30 mm. Forewings have a lighter spot near the center. Hindwings are paler with darker borders; however, a light band occurs at the wing edges. Eggs are flattened half-spheres, white to pinkish, with fine radiating lines from the top center. They are covered by the white, hairs like that give a cotton ball appearance to the egg cluster. The larvae color ranges from bright green to purplish green to blackish. The most common phase is light olive green with a darker strip down the back and a paler stripe along each side. Pupation occurs on or under the soil surface. Pupae are brownish, typical of all close relatives (Mau & Kessing, 2007b).

Life cycle and damage: females lay eggs in clusters that may be several layers deep. They hatch in 5-7 days in warm weather. Young larvae feed gregariously and skeletonize foliage. As they mature, larvae become solitary and eat large irregular holes in foliage. They are a serious defoliator of flower crops, safflower and cotton. The larvae have sixth instars (Capinera, 2010a).

Monitoring: Pheromone traps can be used to detect the presence and intensity of moths.

Economic and action thresholds: action threshold is 0.3 larvae per plant (Capinera, 2010a).

Control: controlling broad leaf weeds and rapid disposal of crop residues after harvesting may reduce population buildup. If the threshold is exceeded application of insecticides is advised. However, insecticides should be applied in early larval stages and achieve good coverage of targeted plants (Mau & Kessing, 2007b).

8. References

Ahmad, S., Ahmad Khan, H., Javed, M., & Rehman, K.-U. (2011). Roost composition and damage assessment of rose-ringed parakeet (*Psittacula krameri*) on maize and sunflower in agro-ecosystem of central Punjab, Pakistan. *International Journal of Agriculture and Biology.* 13. 731-736. ISSN: 1560-8530; ISSN Online: 1814-9596.

Alford, D. V. (2003). Biocontrol of oilseed rape pests: Blackwell Science Ltd. 343 pp. ISBN: 0632054271.

Anitha, V., Wightman, J., & Rogers, D. J. (2005). Management of white grubs (Coleoptera: Scarabaeidae) on groundnut in southern India. *International Journal of Pest Management.* 51: 4. 313-320. http://eprints.icrisat.ac.in/id/eprint/945.

Aslam, M., Razaq, M., Shah, S. A., & Ahmad, F. (2004). Comparative efficacy of different insecticides against sucking pests of cotton. *Journal of Research (Science), Bahauddin Zakariya University, Multan, Pakistan.* 15: 1. 53-58. 1021-1012.

Awasthi, M., Sharma, A., & Singh, R. (2003). Screening of soybean genotypes for resistance against three major insect-pests. Part of Master of Science thesis. National Research Centre for Soybean (ICAR), Khandwa Road, Indore 452 017, MP, India.

Bailey, P. T. (2007). Pests of field crops and pastures: identification and control: Csiro Publishing. 520 pp. 0643067582, 9780643067585.

Baldwin, J. L., Davis, J., & Leonard, R. (2011). Control soybean insect pests. *Louisiana State University Agricultural Center.* from www.lsuagcenter.com

Barlow, V. M. (n.d.). The Bagrada Bug in the Palo Verde Valley. *University of California; Cooperative Extension Riverside County.* from http://cals.arizona.edu/crops/vegetables/advisories/docs/Bagrad_bug_2010_Palumbo.pdf

Bennett, K. V. W., Hutchison, W., Burkness, E., Koch, R., & Potter, B. Bean leaf beetle-snap beans.

Bhubaneswar, A. I. S. (2008). Manual on agricultural production technology: Directorate of Agriculture and Food Production Orissa, Bhubaneswar. 209pp.

Bijlmakers, H. (2008). *Nezara viridula* (Linnaeus). Insect pests of cereals in Ethiopia. *FAO/UNDP Project.* Retrieved 3 September 2011, from http://ethiopia.ipm-info.org/insect_pests_ethiopia/Nezara_viridula.htm

Bligaard, J., Meadow, R., Nielsen, O., & Percy-Smith, A. (1999). Evaluation of felt traps to estimate egg numbers of cabbage root fly, *Delia radicum*, and turnip root fly, *Delia floralis* in commercial crops. *Entomologia Experimentalis et Applicata.* 90: 2. 141-148. 0013-8703. 10.1023/a:1003550229692. http://dx.doi.org/10.1023/A:1003550229692.

Borges, M., Zhang, A., Camp, M. J., & Aldrich, J. R. (2001). Adult diapause morph of the brown stink bug, *Euschistus servus* (Say) (Heteroptera: Pentatomidae). *Neotropical Entomology.* 30: 1. 179-182. 1519-566X.

Bottenberg, H., & Subrahmanyam, P. (1997). Occurrence of groundnut aphids (*Aphis craccivora* Koch) and rosette disease in irrigated dry season groundnut in central Malawi. International Arachis Newsletter. 35-36pp.

Boyd, M. L., & Bailey, W. C. (2005). Soybean Pest Management: Dectes Stem Borer. *University of Missouri Extension.* Retrieved 4 September 2011

Brien, S. O. (2010). Cottonseed bug. Primary industries and fisheries. *Queensland Government.* Retrieved 13 August 2011, from http://www.dpi.qld.gov.au/26_14456.htm

Brien, S. O. (2011). Insect pest management in sunflowers. *Queensland Government.* Retrieved 26 September 2011, from http://www.dpi.qld.gov.au/26_5972.htm

Brier, H. (2010). Soybean looper. *Queensland Government.* Retrieved 7 September 2011, from http://www.dpi.qld.gov.au/26_9672.htm

Brier, H. (2011). Insect pest management in soybeans. *Queensland Government.* Retrieved 13 August 2011, from http://www.dpi.qld.gov.au/26_6076.htm

Brier, H., & McLennan, A. (2006). Poor control of red-banded shield bug in soybeans with currently registered insecticides. Trial conducted March 2006, Childers, Queensland. from http://www.dpi.qld.gov.au

Broatch, J. S., & Vernon, R. S. (1997). Comparison of water traps and sticky traps for monitoring *Delia* spp. (Diptera: Anthomyiidae) in canola. *The Canadian Entomologist.* 129: 5. 979-984. doi:10.4039/Ent129979-5. http://pubs.esc-sec.ca/doi/abs/10.4039/Ent129979-5.

Brooks, J. E., Ejaz, A., & Iftikhar, H. (1988). Characteristics of damage by vertebrate pests to groundnuts in Pakistan. Paper presented at the Thirteenth Vertebrate Pest Conference. University of Nebraska - Lincoln.

CABI. (2008). Datasheets: *Ceutorhynchus assimilis* (cabbage seed weevil). Invasive Species Compendium (Beta). 12 July 2011, from http://www.cabi.org/isc/?compid =5&dsid=12444&loadmodule=datasheet&page=481&site=144

CABI. (2011). Groundnut aphid (*Aphis craccivora*). Empowering farmers, powering research - delivering improved food security. Retrieved 19 August 2011, from http://www. plantwise.org/?dsid=6192&loadmodule=plantwisedatasheet&page=4270&site=234

Capinera, J. L. (2005). Common name: green peach aphid; scientific name: *Myzus persicae* (Sulzer) (Insecta: Hemiptera: Aphididae). *University of Florida*. Retrieved 16 July 2011, from http://entnemdept.ufl.edu/creatures/veg/aphid/green_peach_aphid. htm

Capinera, J. L. (2009). European Earwig *Forficula auricularia* Linnaeus (Insecta: Dermaptera: Forficulidae). Florida: University of Florida, IFAS Extension. (Publication #EENY483).

Capinera, J. L. (2010a). Beet armyworm, *Spodoptera exigua* (Hübner) (Insecta: Lepidoptera: Noctuidae). *University of Florida*. Retrieved 13 August 2011, from http://edis.ifas. ufl.edu/in262

Capinera, J. L. (2010b). Melon thrips: *Thrips palmi* Karny (Insecta: Thysanoptera: Thripidae). Florida Insect Management Guide for vegetables. *University of Florida; and Division of Plant Industry; USDA: Publication Number: EENY-135*. Retrieved 13 August 2011, from http://entnemdept.ufl.edu/creatures/veg/melon_thrips.htm

Chattopadhyay, C., Agrawal, R., Kumar, A., Singh, Y. P., Roy, S. K., Khan, S. A., Bhar, L. M., Chakravarthy, N. V. K., Srivastava, A., Patel, B. S., Srivastava, B., Singh, C. P., & Mehta, S. C. (2005). Forecasting of *Lipaphis erysimi* on oilseed Brassicas in India--a case study. *Crop Protection*. 24: 12. 1042-1053. 0261-2194. http://www. sciencedirect.com/science/article/pii/S0261219405000803.

Chaturvedi, I. (2007). Economic threshold level and monitoring systems: reality and practicaly in *Helicoverpa armigera* (Lepidoptera: noctuidae). *Electronic Journal of Polish Agricultural Universities*. 10: 2. http://www.ejpau.media.pl/volume10/ issue2/art-06.html

Chowdhury, M. (2009). Incidence of saw fly, *Athalia lugens proxima* Klug. as influenced by level of irrigation and fertilizers on mustard. *The Journal of Plant Protection Sciences*. 1: 1. 80-82.

Collins, G., Whitaker, J., Culpepper, S., Roberts, P., & Kemerait, B. (2010). Cotton growth management. The University of Georgia, Cooperative Extension, College of Agricultural and Environmental Sciences. Retrieved 29 August 2011, from www.ugacotton.com.

Daigle, C. J., Boethel, D. J., & Fuxa, J. R. (1988). Parasitoids and pathogens of green cloverworm (Lepidoptera: Noctuidae) on an uncultivated spring host (Vetch, *Vicia* spp.) and a cultivated summer host (Soybean, *Glycine max*). *Environmental Entomology*. 17: 1. 90-96. http://www.ingentaconnect.com/content/esa/envent/ 1988/00000017/00000001/art00016.

De Groote, H. (1997). Potential for mycopesticide use in Africa: socioeconomic analysis. *Paper presented at the LUBILOSA Project Management Committee in Eschborn*. 27 pp.

Derksen, A., Griffiths, K., & Smith, T. R. (2009). 2008-2009 Florida CAPS survey for *Oxycarenus hyalinipennis* (Hemiptera: Oxycarenidae) in South Florida. (No. 2009-05-OH-01). pp: 8.

Dosdall, L. M., Herbut, M. J., & Cowle, N. T. (1994). Susceptibilities of species and cultivars of canola and mustard to infestation by root maggots (*Delia* spp.) (Diptera: Anthomyiidae). *The Canadian Entomologist*. 126: 2. 251-260. doi:10.4039/Ent126251-2. http://pubs.esc-sec.ca/doi/abs/10.4039/Ent126251-2.

El-Sherif, S. I., El-Hawary, I. S., & Mesbah, I. I. (1991). Economic threshold of infestation with the cotton leaf-worm, *Spodoptera littoralis* (Boisd.) (Lepidoptera: Noctuidae) in cotton fields in A.R.E. 3 - Economic injury levels during the different generations. *Arab Journal of Plant Protection*. 9: 2. 116-123. 0255-983X.

Ellsworth, P. C., & Martinez-Carrillo, J. L. (2001). IPM for *Bemisia tabaci*: a case study from North America. *Crop Protection*. 20: 9. 853-869. 0261-2194. http://www.sciencedirect.com/science/article/pii/S0261219401001168.

Evans, A. (2007). Stem boring pests of winter oilseed rape. *Crop and Soil Systems SAC*. ISSN 0142 7695 . ISBN 1 85482 841.

Franzmann, B. (2011). Black sunflower scarab. *Queensland Government*. Retrieved 23 August 2011, from http://www.dpi.qld.gov.au/26_10333.htm

Gagne, R. J., & Woods, W. M. (1988). Native American plant hosts of *Asphondylia websteri* (Diptera: Cecidomyiidae). *Annals of the Entomological Society of America*. 81: 3. 447-448. 0013-8746.

Gangrade, G. A., & Singh, O. P. (1975). Soybean plant response to the attack of *Oberea brevis* Swed. (Col., Cerambycidae). *Zeitschrift für Angewandte Entomologie*. 79: 1-4. 285-290. 1439-0418. 10.1111/j.1439-0418.1975.tb02343.x. http://dx.doi.org/10.1111/j.1439-0418.1975.tb02343.x.

Ghewande, M. P., & Nandagop, A. V. (1997). Integrated pest management in groundnut (*Arachis hypogaea* L.) in India. *Integrated Pest Management Reviews*. 2. 1-15.

Gill, H. K., Capinera, J. L., & McSorley, R. (2008). Common name: lesser cornstalk borer; scientific name: *Elasmopalpus lignosellus* (Zeller) (Insecta: Lepidoptera: Pyralidae) *University of Florida*. Retrieved 23 August 2011, from http://entnemdept.ufl.edu/creatures/field/lesser_cornstalk_borer.htm

Gomez, C., & Mizell, R. F. (2008). Brown stink bug *Euschistus servus* (Say) (Insecta: Hemiptera: Pentatomidae). Florida: University of Florida, IFAS Extension. (EENY 433). 5pp.

Gouge, D., Smith, K., & Henneberry, T. (1998). Field control of pink bollworm *Pectinophora gossypiella* (Saunders) (lepidoptera: Gelechiidae) with entomopathogenic nematodes (Rhabditida: Steinernematidae) and gossyplure in cotton. *Recent Research Developments in Entomology* 2. 55-58.

Gouge, D. H., Way, M., Knutson, A., Cronholm, G., & Patrick, C. (1999). Managing soybean insects. *Texas Agricultural Extension Service, B-1501*.

Graham, C. W., & Gould, H., J. (1980). Cabbage stem weevil (*Ceutorhynchus quadridens*) on spring oilseed rape in Southern England and its control. *Annals of Applied Biology*. 95: 1. 1-10. 1744-7348. 10.1111/j.1744-7348.1980.tb03964.x. http://dx.doi.org/10.1111/j.1744-7348.1980.tb03964.x.

Hadi, B., & Bradshaw, J. (2011). NPIPM: *Pseudoplusia includens* (soybean). Center for Invasive Species and Ecosystem Health at the University of Georgia. Retrieved 18 August

2011, from http://wiki.bugwood.org/NPIPM:Pseudoplusia_includens_%28 soybean%29

Hadi, B., Bradshaw, J., Hunt, T., Knodel, J., Ostlie, K., Whitworth, R. J., Michaud, J. P., & Sloderbeck, P. E. (2011). NPIPM: *Plathypena scabra* (soybean). Center for Invasive Species and Ecosystem Health at the University of Georgia. Retrieved 23 August 2011, from http://wiki.bugwood.org/NPIPM:Plathypena_scabra_%28soybean%29

Hanumantharaya, L., Balikai, R. A., Mallapur, C. P., Venkateshalu, & Kumar, C. J. (2008). Integrated Pest Management Strategies against safflower aphid, *Uroleucon compositae* (Theobald). Paper presented at the 7th International Safflower Conference. Wagga Wagga, NSW, Australia.

Harish, G. (2008). Studies on incidence and management of defoliator pests of soybean. University of Agricultural Sciences, Dharwad. Master of Science thesis. 62pp.

Jasani, H. (2009). Insect pest management in groundnut. *Ministry of Science & Technology, Department of Scientific & Industrial Research, GoI Designed And Developed at Directorate of Instrumentation*. Retrieved 17 August 2011, from http://www.jnkvv. nic.in/ipm%20project/insect-groundnut.html

Jellis, G. (2003). Pest management in cereals and oilseed rape – a guide (pp. 24): The Home-Grown Cereals Authority (HGCA).

Johnson, D. W., Townsend, L. H., Green, J. D., Martin, J. R., Witt, W. W., Hershman, D. E., Murdock, L., & Herbek, J. (2009). Kentucky integrated crop management manual for soybeans. Kentucky: University of Kentucky. 76pp.

Kalita, M. M. (2004). IPM package of practices for field crops. Guwahati. 140 pp.

Kapoor, K. N., Gujrati, J. P., & Gangrade, G. A. (1972). Parasites of *Oberea brevis* (Coleoptera: Lamiidae). *Annals of the Entomological Society of America*. 65: 3. 755-755. http://www.ingentaconnect.com/content/esa/aesa/1972/00000065/00000003/art 00049.

Kathryn, A. B. (2008). Common name: velvetbean caterpillar; scientific name: *Anticarsia gemmatalis* (Hübner) (Insecta: Lepidoptera: Noctuidae) *University of Florida*. Retrieved 13 August 20011, from http://entnemdept.ufl.edu/creatures/field/ velvetbean.htm

Kessing, J. L. M., & Mau, R. F. L. (1991). *Brevicoryne brassicae* (Linnaeus). Crop Knowledge Master. Retrieved 14 July 2011, from http://www.extento.hawaii.edu/kbase/ crop/type/brevicor.htm

Kessing, J. L. M., & Mau, R. F. L. (2007). *Hellula undalis* (Fabricius). *Crop Knowledge Master*. Retrieved 14 August 2011, from http://www.extento.hawaii.edu/kbase/crop/ type/hellula.htm

Khanpara, A. V. (2011). Management of jassid (*Empoasca kerri* Pruthi) in summer groundnut: LAP Lambert Academic Publishing. 88pp. 978-3-8443-9740-6.

Knodel, J., & Olson, D. (2002). Crucifer flea beetle: biology and integrated pest management in canola. *North Dakota State University Extension Service Publication E-1234, North Dakota State University, Fargo, ND, USA*.

Knodel, J. J., Beauzay, P., & Boetel, M. (2010). Field crop insect management guide. North Dakota State University (NDSU) Extension Service. 215 pp. Retrieved 23 August 2011, from http://www.ag.ndsu.edu/.

Knodel, J. J., Charlet, L. D., & Gavloski, J. (n.d.). Integrated pest management of sunflower insect pests in the Northern great plains. (No. E-1457). North Dakota State University and U.S. Department of Agriculture cooperating. 20 pp.

Koehler, P. G. (n.d.). Management of stored grain and peanut pests. Institute of Food and Agricultural Sciences. University of Florida. 6 pp.

Krishworld. (n.d.). Insect pests of crops. Retrieved 25 August 2011, from http://www.krishiworld.com/html/insect_pest_crops13.html

Linker, H. M., Jordan, D., Brandenburg, R., Bailey, J., Herbert, D. A., Phipps, P. M., & Swann, C. W. (n.d.). Integrated pest management peanut scouting manual, North Carolina Cooperative Extension Service: North Carolina State University and Virginia Cooperative Extension Service:Virginia Polytechnic Institute and State University, 27pp,

Lykouressisa, D., Perdikisa, D., Samartzisa, D., Fantinoub, A., & Toutouzasa, S. (2005). Management of the pink bollworm Pectinophora gossypiella (Saunders) (Lepidoptera: Gelechiidae) by mating disruption in cotton fields. Crop Protection. 24: 2. 177-183. doi:10.1016/j.cropro.2004.07.007.

Lynch, R. E. (1990). Resistance in peanut to major arthropod pests. Florida Entomologist. 73: 3. 422.

Mallikarjuna, N., Kranthi, K. R., Jadhav, D. R., Kranthi, S., & Chandra, S. (2004). Influence of foliar chemical compounds on the development of Spodoptera litura (Fab.) in interspecific derivatives of groundnut. JEN. 128: 5. 321-328. 10.1111/j.1439-0418.2004.00834.321-328.

Manyangarirwa, W., Zehnder, G., Mccutcheon, G., Smith, J., Adler, P., & Mphuru, A. (2009). Parasitoids of the diamondback moth on brassicas in Zimbabwe. Paper presented at the African Crop Science Conference Proceedings. 565-570.

Martínez, J. F. (1999). Sesame and Safflower Newsletter, Apartado Córdoba, Spain. Institute of Sustainable Agriculture (IAS), SCIC. Vol. 14.

Martinez, S. S., & Emden, H. F. (2001). Growth disruption, abnormalities and mortality of Spodoptera littoralis (Boisduval) (Lepidoptera: Noctuidae) caused by Azadirachtin. Neotropical Entomology. 30: 1 1519-566X. doi.org/10.1590/S1519-566X2001000100017. http://www.scielo.br/scielo.php?script=sci_arttext&pid=S1519-566X2001000100017.

Mau, R. F., & Kessing, J. L. M. (2007a). Tetranychus cinnabarinus (Boisduval). Crop Knowledge Master. Retrieved 26 August 2011, from http://www.extento.hawaii.edu/kbase/crop/type/t_cinnab.htm

Mau, R. F. L., & Kessing, J. L. M. (2007b). Spodoptera exigua (Hubner). Crop Knowledge Master. Retrieved 26 August 2011, from http://www.extento.hawaii.edu/kbase/crop/type/spodopte.htm

Mau, R. F. L., Kessing, J. L. M., & Hawaii, H. (2007). Trichoplusia ni (Hubner). Retrieved 29 August 2011, from http://www.extento.hawaii.edu/kbase/crop/type/trichopl.htm

Maxwell, E. M., & Fadamiro, H. Y. (2006). Evaluation of several reduced-risk insecticides in combination with an action threshold for managing lepidopteran pests of cole crops in Alabama. Florida Entomologist. 89: 2. 117-126. 0015-4040.

McColl, S. A., Khan, M., & Umina, P. A. (2011). Review of the biology and control of Creontiades dilutus (Stål) (Hemiptera: Miridae). Australian Journal of Entomology. 50: 2. 107-117. 1440-6055. 10.1111/j.1440-6055.2010.00800.x. http://dx.doi.org/10.1111/j.1440-6055.2010.00800.x.

Menon, M., & Jayaraman, N. (2002). Bt Cotton: seeds of discontent. Agriculture and Biotechnology, India Resource Center. http://www.IndiaResource.org.

Michaud, J. P., Whitworth, R. J., & Davis, H. N. (2011). Sunflower insect management. Kansas State University.

Micic, S., Henry, K., & Horne, P. (2007). Identification and control of pest slugs and snails for broadacre crops in Western Australia. *Bulletin 4713*. 16 pp. ISSN: 1833-7236.

Mossler, M., & Aerts, M. J. (2009). Florida Crop/pest management profiles: peanuts. *University of Florida*. Retrieved 23 June 2011

Mundel, H. H., & Johnson, D. L. (1987). Safflower susceptibility and response to feeding by grasshoppers. *Annals of Applied Biology*. 111: 1. 203-212. 1744-7348.

Murchie, A., Williams, I. H., & Perry, J. (1999). Edge distributions of *Ceutorhynchus assimilis* (Paykull) and its parasitoid *Trichomalus perfectus* (Walker) in a crop of winter oilseed rape (*Brassica napus* L.). *Biocontrol*. 44. 379-390.

Murray, D. (2010). False wireworm. *Queensland Government*. Retrieved 4 September 2011, from http://www.dpi.qld.gov.au/26_8135.htm

Nandagopal, V., & Reddy, P. S. (1987). Outbreaks of *Balclutha hortensis* Lindb - a new pest species of groundnut. *Journal of Oilseeds Research*. 4: 2. 300. 0970-2776.

Natarajan, K. (2007). Management of agriculturally important sucking pests of cotton. Central Institute for Cotton Research, Regional Station, Coimbatore. 80-83 pp.

Navarajan, P. A. V. (2007). Insect pests and their management. (No. 110012). 68pp.

Nielsen, C., & Hajek, A. E. (2005). Control of invasive soybean aphid, *Aphis glycines* (Hemiptera: Aphididae), populations by existing natural enemies in New York State, with emphasis on entomopathogenic fungi. *Environmental Entomology*. 34: 5. 1036-1047. 0046-225X. 10.1603/0046-225x(2005)034[1036:coisaa]2.0.co;2. http://dx.doi.org/10.1603/0046-225X(2005)034[1036:COISAA]2.0.CO;2.

NSW. (2001). Subfamily Deltocephalinae: Tribe Macrostelini. New South Wales Government Industry and Investment NSW Agricultural Scientific Collections Unit. Retrieved 23 August 2011, from http://www1.dpi.nsw.gov.au/keys/leafhop/deltocephalinae/macrostelini.htm

Ohnesorge, B., & Rapp, G. (1986). Monitoring *Bemisia tabaci*: A review. *Agriculture, Ecosystems & Environment*. 17: 1-2. 21-27. 0167-8809. http://www.sciencedirect.com/science/article/pii/016788098690023X.

Oliveira, M. R. V., Henneberry, T. J., & Anderson, P. (2001). History, current status, and collaborative research projects for *Bemisia tabaci*. *Crop Protection*. 20: 9. 709-723. 0261-2194. http://www.sciencedirect.com/science/article/pii/S0261219401001089.

Olson, D. M., Ruberson, J. R., Zeilinger, A. R., & Andow, D. A. (2011). Colonization preference of *Euschistus servus* and *Nezara viridula* in transgenic cotton varieties, peanut, and soybean. *Entomologia Experimentalis et Applicata*. 139: 2. 161-169. 1570-7458. 10.1111/j.1570-7458.2011.01116.x. http://dx.doi.org/10.1111/j.1570-7458.2011.01116.x.

OMAFRA. (2011). Insects and pests of field crops: soybean insects and pests. Agronomy Guide for Field Crops. *Ontario Ministry of Agriculture Food and Rural Affairs*. Retrieved 26 August 2011, from http://www.omafra.gov.on.ca/english/crops/pub811/13soybean.htm

Pande, A. K., Mahajan, S. V., Munde, A. T., & Telang, S. M. (2000). Evaluation of insecticides against major pests of soybean (*Glycine max* L.). *Journal of Soils and Crops*. 10: 2. 244-247. 0971-2836.

Parker, B. L., Talekar, N. S., & Skinner, M. (2001). Soybean insect pests: armyworm, *Spodoptera litura*. Field Guide: Insect Pests of Selected Vegetables in Tropical and Subtropical Asia. from http://www.avrdc.org/LC/soybean/armyworm.html

Parshad, V. R. (1999). Rodent control in India. *Integrated Pest Management Reviews.* 4. 97-126.

Patil, R., & Hegde, R. (2010). Biorational approaches for the management of lepidopteran pests of soybean. *Karnataka Journal of Agricultural Sciences.* 22: 3.

Pavela, R., & Kazda, J. (2007). Influence of application term on effectiveness of some insecticides against brassica pod midge (Dasineura brassicae Winn.). *Plant Protection Science-UZPI.* 43: 1212-2580.

Pineda, S., Schneider, M., Smagghe, G., Martinez, A. M., Estal, P. D., Uela, E. V., Valle, J., & Budia, F. (2007). Lethal and sublethal effects of methoxyfenozide and spinosad on *Spodoptera littoralis* (Lepidoptera: Noctuidae). *Journal of Economic Entomology.* 100: 3. 773-780.

Putnam, D. H., Oplinger, E. S., Hicks, D. R., Durgan, B. R., Noetzel, D. M., Meronuck, R. A., Doll, J. D., & Schulte, E. E. (1990). Sunflower: Alternative field crops manual. Retrieved 30 July 2011, from http://www.hort.purdue.edu/newcrop/afcm/sunflower.html

Qureshi, Z. A., & Ahmed, N. (1991). Monitoring seasonal population fluctuation of spotted and spiny bollworms by synthetic sex pheromones and its relationships to boll infestation in cotton1. *Journal of Applied Entomology.* 112: 1-5. 171-175. 1439-0418. 10.1111/j.1439-0418.1991.tb01043.x.
http://dx.doi.org/10.1111/j.1439-0418.1991.tb01043.x.

RADA. (2010). Biology, behaviour and management of the gall midge complex on hot peppers. Rural Agricultural Development Authority. *Division of Technology, Training & Technical Information, Rural Agricultural Development Authority.* Retrieved 13 August 2011, from http://www.rada.gov.jm/get_articles.php?ai=51

Rana, Z. (n.d.). *Psittacula krameri* [Scopoli 1769]. Retrieved 5 September 2011, from http://home.wanadoo.nl/psittaculaworld/Species/P-krameri.htm

Ranga Rao, G., Rameshwar Rao, V., & Nigam, S. (2010). Post-harvest insect pests of groundnut and their management [Electronic Version], *Bulletin No. 84,* International Crops Research Institute for the Semi-Arid Tropics, 23pp. Retrieved 23 August 2011,

Rao, G. R., Desai, S., Rupela, O., Krishnappa, K., & Wani, S. P. (2007). Best-best options for integrated watershed management: integrated pest management options for better crop production. Paper presented at the Proceedings of the Comprehensive Assessment of Watershed Programs in India.International Crops Research Institute for the Semi-Arid Tropics. 314pp.

Rueda, A. (1995). Onion Thrips. *Cornell University.* Retrieved 13 August 2011, from http://web.entomology.cornell.edu/shelton/veg-insects-global/english/thrips.html

Ruther, J., & Thiemann, K. (1997). Response of the pollen beetle *Meligethes aeneus* to volatiles emitted by intact plants and conspecifics. *Entomologia Experimentalis et Applicata.* 84: 2. 183-188. 0013-8703. 10.1023/a:1003040318376. http://dx.doi.org/10.1023/A:1003040318376.

Sabesh, M. (2004). Integrated Pest Management practice for Cotton. *Ministry of Agriculture, Department of Agriculture & Cooperation, Directorate of Plant Protection, Quarantine & Storage, Government of India.* Retrieved 7 September 2011, from http://www.cicr.org.in/PDF/ipm.pdf

Saeidi, K., & Nur Azura, A. (2011). Biology of *Acanthiophilus helianthi* Rossi (Diptera: Tephritidae): a safflower pest of soutern Iran. Paper presented at the VIIth Arthropods, 18-23 September, 2011. Białka Tatrzańska, Poland. p: 57.

Sahayaraj, K., & Martin, P. (2003). Assessment of *Rhynocoris marginatus* (Fab.) (Hemiptera: Reduviidae) as augmented control in groundnut pests. *Journal of Central European Agriculture*. 4: 2. 103-110. 1332-9049.

Salama, H. S. (1983). Cotton-pest management in Egypt. *Crop Protection*. 2: 2. 183-191. 0261-2194. http://www.sciencedirect.com/science/article/pii/0261219483900431.

Sanchez-Arroyo, H. (2009). Common name: Mexican bean beetle scientific name: *Epilachna varivestis* Mulsant (Insecta: Coleoptera: Coccinellidae). *University of Florida*. Retrieved 5 September 2011, from http://entnemdept.ufl.edu/creatures/veg/bean/mexican_bean_beetle.htm

Santhosh, M. (2008). Evaluation of ITK components against major pests of soybean (*Glycine max* (L.) Merrill). University of Agricultural Sciences, Dharwad. Master of Science thesis. 100pp.

Savajji, K. (2006). Biology and management of soybean stem fly *Melanagromyza sojae* (Zehntner) (Diptera: Agromyzidae). University of Agricultural Sciences. Master of Science thesis. 58pp.

Smith, T. R., & Brambila, J. (2008). A major pest of cotton, *Oxycarenus hyalinipennis* (Heteroptera: Oxycarenidae) in the Bahamas. *Florida Entomologist*. 91: 3. 479-482. ISSN: 0015-4040.

TIFP. (n.d.-a). Insect pest management in groundnut. *Ministry of Science & Technology, Department of Scientific & Industrial Research, GoI Designed And Developed at Directorate of Instrumentation*. Retrieved 17 August 2011, from http://www.jnkvv.nic.in/ipm%20project/insect-groundnut.html

TIFP. (n.d.-b). Insect pest management in mustard. Integrated Pest Management. *Ministry of Science & Technology, Dpt. of Scientific & Industrial Research, GoI* Retrieved 26 August 2011, from http://www.jnkvv.nic.in/ipm%20project/insect-mustard.html

TIFP. (n.d.-c). Insect pest management in sunflower *Ministry of Science & Technology, Department of Scientific & Industrial Research, GoI Designed And Developed at Directorate of Instrumentation*. Retrieved 17 August 2011, from http://www.jnkvv.nic.in/ipm%20project/insect-sunflower.html

TNAU. (n.d.-a). Crop insect pest: oil seeds: pest of safflower. *Tamilnadu Agricultural University, Coimbatore*. Retrieved 8 August 2011, from http://agritech.tnau.ac.in/crop_protection/crop_prot_crop_insect_oil_safflower.html

TNAU. (n.d.-b). Crop insect pest: oil seeds: pest of sunflower. *Tamilnadu Agricultural University, Coimbatore*. . Retrieved 23 September 2011

TNAU. (n.d.-c). Crop insect pest: pulses: Pest of soybean. *Tamilnadu Agricultural University, Coimbatore*. from http://agritech.tnau.ac.in/crop_protection/crop_prot_crop_insect_pulsh_soyabean.html

Udikeri, S. S., Patil, S. B., Krishna naik, L., Rachappa, V., Nimbaland, F., & Guruprasad, G. S. (2007). Poncho 600 FS – A new seed dressing formulation for sucking pest management in cotton. *Karnataka Journal of Agricultural Science*. 20: 1. 51-53.

Umeh, V. C., Youm, O., & Waliyar, F. (2001). Mini Review - soil pests of groundnut in sub-Saharan Africa, International Centre for Insect Physiology and Ecology (ICIPE), Kenya, from http://hdl.handle.net/1807/162

USDA. (2010). New pest response guidelines: cotton seed bug. Emergency and Domestic Programs. United States Department of Agriculture, Animal and Plant Health Inspection Service and Cooperating State Departments of Agriculture. 114 pp. Retrieved 23 August 2011.

Vaamonde, C. L. (2006). *Spodoptera littoralis*. Delivering Alien Invasive Species Inventories for Europe. from http://www.europe-aliens.org/pdf/Spodoptera_littoralis.pdf

Vennila, S. (n.d.). Cotton pests, predators and parasitoids: Descriptions and seasonal dynamics. Retrieved 13 August 2011, from www.cicr.gov.in

Viteri, D., Cabrera, I., & Estévez de Jensen, C. (2010). Identification and abundance of thrips species on soybean in Puerto Rico. *International Journal of Tropical Insect Science*. 30: 01. 57-60. 1742-7592.

Wikipedia. (2010). Groundnut. The free encyclopedia. *Wikimedia Foundation Inc.* Retrieved 30 July 2011, from http://en.wikipedia.org/wiki/Groundnut

Wikipedia. (2011a). Cotton. The free encyclopedia. *Wikimedia Foundation Inc.* Retrieved 28 July 2011, from http://en.wikipedia.org/wiki/Cotton

Wikipedia. (2011b). *Delia radicum*. The free encyclopedia. *Wikimedia Foundation Inc.* Retrieved 30 August 2011, from http://en.wikipedia.org/wiki/Delia_radicum

Wikipedia. (2011c). Green stink bug. The free encyclopedia. *Wikimedia Foundation Inc.* Retrieved 30 August 2011, from http://en.wikipedia.org/wiki/Green_stink_bug

Wikipedia. (2011d). *Scirtothrips dorsalis*. The free encyclopedia. *Wikimedia Foundation Inc.* Retrieved 13 August 2011, from http://en.wikipedia.org/wiki/Scirtothripsdorsalis

Wikipedia. (2011e). Soybean. The free encyclopedia. *Wikimedia Foundation Inc.* Retrieved 30 July 2011, from http://en.wikipedia.org/wiki/Soybean

Wilson, L. (2011). Cotton pests thrive in a wet season. Cotton Pest Management Guide. Retrieved 13 August 2011, from www. cottoncrc.org.au/

Winfield, A. (1961). Observations on the biology and control of the cabbage stem weevil, *Ceutorhynchus quadridens* (Panz.), on Trowse mustard (*Brassica juncea*). *Bulletin of Entomological Research*. 52: 03. 589-600. 1475-2670.

Yaku, A., Walter, G., & Najar Rodriguez, A. (2007). Thrips see red–flower colour and the host relationships of a polyphagous anthophilic thrips. *Ecological Entomology*. 32: 5. 527-535. 1365-2311.

Yukawa, J., Uechi, N., Horikiri, M., & Tuda, M. (2003). Description of the soybean pod gall midge, *Asphondylia yushimai* sp. n. (Diptera: Cecidomyiidae), a major pest of soybean and findings of host alternation. *Bulletin of Entomological Research*. 93. 73-86. DOI: 10.1079/BER2002218.

Yvette, C., & Jensen, B. (2009). Cotton pest management guide. Cotton CRC Extension Team. State of New South Wales through Department of Industry and Investment (Industry & Investment NSW). 154pp. Retrieved 2009-2010.

Zinni, Y. (2011). Scientific Names for oregon's grasshoppers. Retrieved 23 August 2011, from http://www.ehow.com/info_8409456_scientific-names-oregons-grasshoppers.html

6

Effect of Seed-Placed Ammonium Sulfate and Monoammonium Phosphate on Germination, Emergence and Early Plant Biomass Production of Brassicae Oilseed Crops

P. Qian[1], R. Urton[1], J. J. Schoenau[1], T. King[1], C. Fatteicher[1] and C. Grant[2]

[1]*University of Saskatchewan, Saskatoon, SK,*
[2]*Agriculture and Agri-Food Canada, Brandon, MB,*
Canada

1. Introduction

Seed-placed fertilization, in which fertilizer is placed in the soil in the same furrow as the seed at the time of planting is a common approach to supplying crop nutrients, as this gives newly emerged seedlings early access to nutrients. This placement strategy is noted to be effective for phosphorus fertilizer due to its low mobility in the soil in early spring (Miller et al., 1971; Harapiak, 2006). In the prairies of Canada, cool soil temperatures in the spring at seeding can especially restrict early root growth and access to phosphorus. Therefore it is important for annual crops to be able to access P early in their growth cycle by placing P fertilizer where the roots of the seedling can readily access it, such as in the seed-row. This is the reason why seed-placed phosphorus fertilization normally achieves better response than surface-applied with incorporation. Lauzon and Miller (1997) reported that early season corn and soybean shoot-P concentrations are increased with increasing soil test P and were increased with seed-placed P regardless of soil test P level. However, the germination and emergence of crop seeds can be reduced by seed-placed phosphate fertilizer, as some crop seeds are especially sensitive to the salt effect of fertilizers (Qian and Schoenau, 2010).

In western Canada, canola is a major crop and approximately 4.5 million hectares of agricultural land are under *Brassica* oilseed crop production (Malhi et al., 2007). Canola-quality *B. juncea*, *B. carinata*, and oilseed *Camelina sativa* are also being developed as alternative oilseed crops that are better adapted to areas with periods of hot, dry conditions in western Canada. Tolerance of *Brassica* crops to seed-row application of nutrients is low when compared to many other crops, and emergence differences have been observed between open-pollinated and hybrid cultivars (Brandt et al., 2007). Differences have also been noted in the tolerance and responsiveness of yellow- and black-seeded canola cultivars to seed-placed P (Grant, 2008). Qian and Schoenau (2010) reported that canola seed, in general, is sensitive when the seed-placed rate of P is above 30 kg P_2O_5 ha[-1]. The development and adoption of maximum safe rates of fertilizer with seed that avoid significant reductions in crop emergence is important for achieving maximum benefit from the fertilizer by producers.

Brassica crops like canola have higher demand for S than cereal crops. Canola requires about twice the amount of sulfur as cereal crops do, and canola frequently responds to fertilizer S addition in the drier rain-fed cropping areas of the prairie provinces of Canada, where no-till cropping systems are predominant (Canola Council of Canada 2010). A balance in availability of N, S and P for canola is important for both canola yield and quality (Janzen and Bettany, 1984; Jackson, 2000; Karamanos et al., 2005, Qian and Schoenau, 2007). In Canada, the main S fertilizer used is ammonium sulphate (21-0-0-24). Ammonium sulphate, both as prills and as fines, has become a popular source of N and S fertilizer for canola growers. Given the increased prevalence and availability of ammonium sulfate originating as by-products of industrial processes such as flue gas scrubbing, it seems likely that ammonium sulfate will be cost-effective and its use is likely to increase in the future. However, to date there has been little attention given to the tolerance of Brassica crops like canola to seed-row placed ammonium sulfate, both alone and in combination with starter phosphorus. Also, ammonium sulfate has a higher salt index than monoammonium phosphate and can produce significant amounts of free ammonia, leading to the possibility of both osmotic damage and direct ammonia toxicity at high application rates (Follett et al., 1981). Therefore, providing guidelines for maximum safe rates of fertilizer P and S with the seed is essential for achieving maximum benefit from seed-row placement of fertilizer. Current guidelines state that the amount of S that can be safely placed with the seed of canola as ammonium sulfate should be based on N guidelines, and no specific information is provided on tolerance and response to combinations of ammonium sulfate and starter monoammonium phosphate. Nevertheless, existing guidelines were developed some time ago based on seeding equipment with high disturbance and high seed-bed utilization. The trend today is towards low disturbance opener systems. Growers have questioned how much ammonium sulfate can be safely placed in the seed-row along with P as a starter blend for canola, especially with low soil disturbance opener configurations currently used. The objectives of this chapter are to: (1) determine the effects of seed-placed ammonium sulfate (AS) and monoammonium phosphate (MAP) fertilizer applied at different rates on seedling germination and emergence as well as early season dry matter yield under controlled-environmental conditions; and (2) determine if *Brassica* oilseed crops/cultivars differ in their response to seed placed S and P fertilizer. To fully achieve these objectives, experimental works were conducted as described below, and results and conclusions are drawn.

2. Methods and materials

2.1 Soil, fertilizer and crop seeds

The soil selected for the study is a Brown Chernozem (US classification: Aridic Haploboroll), loamy textured soil from a long-term alfalfa field (legal location: SW31, Township 20, Range 3, W of 3rd Meridian) in southwestern Saskatchewan, Canada. The soil represents a common soil type (Haverhill association) found in the southern prairies of Saskatchewan. As there was no history of herbicide application, there was no concern with herbicide residue carryover that could affect germination. The soil was collected from the 0-15 cm depth in the fall of 2010 from a field that has been in continuous alfalfa for ten years. The main soil characteristics and nutrient levels are shown in Table 1.

Soil	pH	Organic C %	Texture	NO$_3$-N	Available/Extractable		
					P	K	SO$_4$-S
					——————mgkg^{-1}——————		
Haverhill	8.0	2.3	loam	45.7	12.3	693	24.0

Table 1. Some characteristics of the soil used for seed-placed S and P evaluation.

After collection, the soil was mixed thoroughly in a soil mixer and stored in field-moist condition before use. For measuring basic soil properties, a sample was collected from the mixed soil, and then air-dried, crushed, and passed through a 2-mm sieve and stored at room temperature. Texture was estimated by hand. Electrical conductivity (EC) and pH were measured using 1:1 soil:water suspension. Organic C was measured using an

Species and Cultivar	Type	Herbicide system	Year of release	Maturity
B. napus				
5440	HYB	LL	2007	Mid-Late
5525	OP	CL	2009	Mid-Late
45H26	HYB	RR	2006	Mid
v1037*	HYB	RR	2007	Early-Mid
v1040*	HYB	RR	2010	Mid
74P00	OP	LL	2006	Early-Mid
H.E.A.R*	OP	RR		
B. juncea				
Dahinda	OP	Conv.	2004	Late
Xceed 8571	OP	CL	2008	Early
B. rapa				
ACS- C7	SYN-OP	Conv.	2001	Early
Camelina sativa				
Calena	OP	Conv.	N.A.	Early
B. carinata	OP	Conv.	N.A.	Late

*-specialty oil; HYB-Hybrid; OP-open-pollinated; SYN-synthetic; Conv.-conventional
LL-Liberty Link; RR-Roundup Ready; CL-Clearfield

Table 2. Selected *Brassica* species/cultivars used in this study.

automated combustion LECO carbon analyzer. Soil available P and K were extracted by modified Kelowna method (Qian et al., 1994). Soil available N was calculated as the sum of NO_3-N + NH_4-N. Both forms of inorganic N were extracted with 2 M KCl (Keeney and Nelson, 1982) and measured by automated colorimetry. Soil available S was extracted with 0.001 M $CaCl_2$ solution and measured by automated colormetry (Qian et al., 1992). The fertilizer phosphorus source used was conventional commercial fertilizer grade granular monoammonium phosphate MAP (12-51-0). The sulfate source used was fertilizer grade prilled ammonium sulfate $(NH4)_2SO_4$ (21-0-0-24). The *Brassica* species/cultivars selected for this study are listed in Table 2. Among the 12 selected cultivars, four different categories of canola were included: open pollinated (OP) and hybrid *napus* (also termed "Argentine" canola) (HYB), *rapa* (also termed "Polish" canola), and *juncea* (sometimes termed "mustard canola"). Although not currently grown on a large commercial scale, *B. carinata* and *C. sativa* are currently under development as new oilseed crops with similar attributes to canola and were therefore included in the study as well.

2.2 Laboratory study

Plants were grown in plastic trays (52cm x 26cm x 6cm) containing 5.4 kg of uniformly mixed, air-dried soil at 20 degrees C. The soil in each tray was levelled to a height of 5cm over the individual rows and packed. Six 20cm x 2.50cm x 1.25cm seed-rows were created in the trays using a seeding tool. The crops were seeded using a seed quantity per unit area in the row that is equivalent to a seeding rate of approximately 10 kg ha^{-1}. The seeding tool used creates a seed bed utilization of approximately 15%, in which 15% of the total seed bed area is used for placement of seed and fertilizer in rows. This seed bed utilization is typical of the wide row spacing, narrow opener seeding tool configurations commonly used for seeding oilseeds in the northern Great Plains today. Sixteen seeds were seeded in each row at a uniform depth of 1.25cm.

The fertilizer was passed through a 2mm sieve to provide uniform granule size and then spread uniformly down the seed-row with the seed. During germination, trays were kept under constant light and regularly watered to maintain soil moisture at 100% water holding capacity for the first two days and ensure germination, and then reduced to 80% of field capacity to maintain soil moisture for seedling growth. Emergence counts were taken every two days after seeding until 14 days after seeding (DAS) when plant counts were constant and no additional emergence was observed. Plants were harvested 14 DAS. Plant biomass samples were washed in deionized water after cutting at the soil surface and oven-dried at 45°C for 3 d to a constant weight.

Six treatments consisting of an unfertilized control and five rates of seed-placed S (10, 20, 30, 40 and 50 kg S ha^{-1}), applied as ammonium sulfate (AS 21-0-0-24) were applied in combination with three rates of seed-placed P_2O_5 (0, 15 and 30 kg P_2O_5 ha^{-1}), applied as monoammonium phosphate (MAP 12-51-0).

The *B.* species/cultivars were designated as main plots, S rates as subplots and P rates as sub-subplots within the trays and were arranged in a randomized complete block design with four replications. Nitrogen was applied at 10, 20, 30, 40, 50 kg N ha^{-1}with the ammonium sulfate alone, and with 15 kg P_2O_5 ha^{-1} as well as 30 kg P_2O_5 ha^{-1} in the seed-row.

2.3 Statistical analysis

Main and interaction effects of *Brassica* cultivars and ammonium sulfate, alone and in combination with monoammonium phosphate, were determined from analysis of variance (ANOVA) using the GLM procedure in SAS (SAS Institute, 2008). Least significant difference (LSD 0.05) was used to determine significant differences between treatment means.

3. Results and discussion

3.1 Effect of seed-row placed ammonium sulfate on brassicae emergence and early season biomass

Generally, there was no significant difference in percent emergence up to 20-30 kg S ha^{-1} when ammonium sulfate was applied alone with the exception of *Camelina sativa*. *Camelina sativa* appeared more sensitive to AS placed in the seed-row than the *Brassica* crops (Table 3). At rates of 30 kg S ha^{-1}, only *Invigor 5440* and *B. Juncea Dahinda* showed inhibition effect. When application rate approached 40 kg S ha^{-1}, the majority of cultivars were affected. The cultivars *HEAR, 5525 Clearfield* and *74P00LL* were the least sensitive to seed-row placement of AS, possibly reflecting greater seed vigor. In general, percentage emergence that dropped below 80% was observed in some *Brassica* cultivars at rates of 30 kg S ha^{-1}, and more so at the rates of 40 kg S ha^{-1} and above (Table 3).

High oleic, low linolenic (HOLL) (*v1037; v1040*) were also less sensitive to seed-row placement of AS; however, they were not as resistant as high erucic acid rapeseed (H.E.A.R). The number of days to first seedling emergence increased with increasing S rate. Time to seedling emergence from seeding was approximately 5-7 d. Emergence was generally 1 to 2 d longer in the high S rate treatments. There was a close relationship between seed size and seed coat color and a decrease in percent emergence, with the yellow-seeded and small seeded cultivars slightly more prone to reduced emergence with seed-placed AS. In most cases, the larger the seed size, the better the vigour. The more vigour, the better the seed/seedling is able to cope with early stresses and survive (Canada Canola Council 2005). High rates of AS reduced the seedling emergence, early seedling growth and increased the time to maximum emergence of seeds. There was an increase in the number of abnormal seedlings observed in the higher S treatments due to seedling injury, which can be attributed to the salt effect of the fertilizer playing a major role in this case (Follett et al., 1981).

Seedling biomass yield (mg / pot) at 14 days of growth was not significantly affected up to rates of 20 – 30 kg S ha^{-1} (Table 4). At higher rates in which significant emergence reduction was observed, seedling biomass was also reduced. As Hall (2007) indicated, canola is much more sensitive to seed-place fertilizer than corn or cereals.

3.2 Effect of combinations of ammonium sulfate and monoammonium phosphate on emergence and early season biomass

Addition of 15 to 30 kg P$_2$O$_5$ ha^{-1} along with the S application led to more injury represented by reduced seed germination/emergence and early growth of seedlings, than S alone at corresponding rates. Canola usually shows injury response to MAP alone at rates of 30 kg P$_2$O$_5$ ha^{-1} and higher (Qian and Schoenau 2010). However, decreases from addition of the

Cultivar	S (Kg S ha⁻¹)						S+15P₂O₅ (Kg S ha⁻¹)						S+30P₂O₅ (Kg S ha⁻¹)					
	0	10	20	30	40	50	0	10	20	30	40	50	0	10	20	30	40	50
B. napus																		
Invigor 5440	100a	97a	92a	68b	57b	32c	100a	102a	82ab	76b	49c	24d	100a	100a	81a	81a	47b	32b
Victory 1037	100ab	99ab	93a	84abc	84bc	61c	100a	97a	88ab	81abc	79bc	67c	100a	89a	81a	79ab	69ab	63b
Victory 1040	100a	102a	106a	99a	65b	56b	100a	100a	93a	82a	60b	49b	100a	92a	85a	83a	58b	49b
HEAR 5525	100a	95a	99a	95a	86a	65b	100a	98ab	100a	100a	84bc	72c	100a	90a	88a	88a	63b	61b
Clearfield	100a	99a	93a	84a	84a	61b	100a	97ab	88ab	81abc	79bc	61c	100a	89ab	86abc	79abc	69bc	63c
74P00 LL	100a	98a	94a	80a	80a	24b	100a	83a	87a	59b	41b	37b	100a	56b	54b	42b	30b	34b
45H26 RR	100a	102a	100a	100a	84b	76b	100a	97a	94a	83b	84b	78b	100a	97ab	95ab	92abc	88bc	83c
B. rapa																		
ASC-7 B. rapa	100a	88ab	100a	89ab	63bc	40c	100a	65b	49bc	47bc	37c	29c	100a	61b	55b	55b	21c	27c
B. juncea																		
Xceed	100ab	100ab	109a	83abc	72bc	66c	100a	80b	90ab	80ab	72bc	50c	100a	89a	81a	94a	57b	53b
Dahinda	100a	89ab	103a	73bc	53cd	31d	100a	93a	70b	70b	40c	7d	100a	97a	59b	47b	22c	9c
Camelina sativa	100a	91ab	71bc	63c	23d	17d	100a	103a	63b	69b	50bc	25c	100a	47b	38bc	27cd	15de	9e
B. carinata	100a	99a	85a	99a	57b	46b	100a	98a	82ab	95a	67b	63b	100a	69b	62bc	59bc	36c	43bc

Numbers in a column followed by the same letter are not significantly different at P<0.05.

Table 3. Mean germination (percentage of unfertilized control) of *Brassica* species with seed-row applied AS and MAP.

	Plant tissue biomass (mg)																	
	S Kg S ha⁻¹						S+15P_2O_5 Kg S ha⁻¹						S+30P_2O_5 Kg S ha⁻¹					
	0	10	20	30	40	50	0	10	20	30	40	50	0	10	20	30	40	50
B. napa																		
Invigor 5440	78a	84a	76a	52b	44b	23c	64ab	71a	60ab	48bc	33c	14d	66a	68a	59a	53a	28b	18b
Victory 1037	72ab	78ab	90a	64bc	48cd	41d	79a	78a	79a	71a	54b	51b	68ab	70ab	76a	60bc	49cd	42d
Victory 1040	67a	67a	69a	69a	39b	36b	71ab	73a	64ab	55b	37c	27c	67a	61a	57a	56a	34b	25b
HEAR	55ab	53ab	56a	51ab	45b	33c	52ab	52a	56a	54a	43bc	36c	55a	52a	53a	50a	36b	34b
5525 Clearfield	55ab	59a	55ab	50b	47b	30c	64ab	72a	62ab	50bc	45cd	36d	67a	67a	57ab	53bc	40cd	37d
74P00 LL	52a	48a	37b	37b	34b	9c	49a	37b	31bc	23cd	14d	17d	52a	25b	26b	19b	15b	13b
45H26 RR	111a	117a	115a	104a	85b	72b	119a	105ab	93bc	82c	83c	64d	114a	108a	99ab	89b	83bc	73c
B. rapa																		
ASC-C7	18ab	20a	24a	17ab	11bc	6c	30a	16b	15bc	8c	10bc	9bc	30ab	22ab	37a	11ab	6ab	5b
B.juncea																		
8571	51a	53a	54a	44ab	32bc	28c	55a	42b	43b	34b	32b	22c	50a	46ab	34bc	46ab	25c	23c
Dahinda	24a	22a	26a	20a	10b	7b	22a	22a	16ab	18a	9b	1c	25ab	29a	16bc	8cd	4d	2d
Camelina sativa	11a	11a	7ab	6b	4bc	2c	11a	10a	6b	6b	4b	2b	11a	7b	5c	4cd	2de	1e
B. carinata	121a	105ab	69cd	85bc	44de	33e	131a	121ab	78cd	93bc	64cd	57d	133a	65b	51bc	36cd	27cd	23d

Numbers in a column followed by the same letter are not significantly different at P<0.05.

Table 4. Mean plant tissue biomass (mg) of 12 oilseed cultivars with seed-row applied AS and MAP.

MAP were typically smaller with the *B. napus* and *B. carinata*. The *B. rapa* and *B. juncea cv. Dahinda* appeared more sensitive to the addition of the P along with S. Among these, the *B. rapa cv. ACS- C7* was particular sensitive to P addition for both rates (15 and 30 kg P_2O_5 ha[-1]). When AS alone was used, the injury in this cultivar was observed at AS rate of 40 kg S ha[-1]; but the reduction of emergence with both rates of P addition was observed at AS rate of 10 kg S ha[-1] (Table 3). For the two cultivars of *B. juncea*, the *Dahinda* cultivar was less tolerant to AS with MAP than *Xceed 8571* (Table 3). The seed sample from which the *Dahinda* was taken was two years old, which may have affected seed vigour and reduced germination. Seeds with lower vigour result in greater reduction in emergence (Canada Canola Council 2005).

Overall, while the high rate (30 kg P_2O_5 ha[-1]) sometimes reduced emergence and biomass production compared to the low rate (15 kg P_2O_5 ha[-1]), often the reductions were not large (Table 3) for most cultivars tested. This agrees with earlier findings that the adverse effect of MAP-P on seed germination and biomass of canola became pronounced at rates over 30 kg P_2O_5 ha[-1] (Qian and Schoenau, 2010).

4. Conclusion

Rates of seed-row placed ammonium sulfate above 20–30 kg S ha[-1] were associated with significant reductions in emergence and biomass of many *Brassica* species/cultivars. Addition of 15 – 30 kg P_2O_5 ha[-1] MAP along with AS often caused further reductions in emergence and biomass, although these were generally not large with *B. napus* cultivars. Differences in tolerance to seed row placed S and P were observed among cultivars. The cultivar 45H26 RR was the most tolerant of cultivars tested, while the most sensitive to seed-row placed S and S+P were *B. rapa, B. juncea* cv. Dahinda, and *Camelina sativa*. Further study is required in the field to establish whether seeds grown under different growing conditions and soil types have similar responses.

5. Acknowledgements

The financial support of Canola Council of Canada is greatly appreciated. *Brassica* and *Camelina* cultivar seeds for this study were obtained from Sherrilyn Phelps, Crop Development Specialist, Saskatchewan Ministry of Agriculture. Soil analysis was provided by ALS Laboratory Group, Saskatoon SK.

6. References

Brandt, S.A., Malhi, S.S., Ulrich, D.J., Kutcher, H.R., and Johnston, A.M. 2007. Seeding rate, fertilizer level and disease management effects on hybrid versus open pollinated canola (*Brassica napus* L.). Canadian Journal of Plant Science, 87(2), pp. 255-266.

Canada Canola Council. 2005, Factors Affecting Canola Survival from Seeding to 21 Days after Emergence, Canola@Fact April 21, 2005, Winnipeg, MB Canada R3B 0T6. http://www.derekerbseeds.com/pdf/agronomy/canola/planting_factorsaffecting canolasurvivalfromseedingto21daysafteremergence.pdf

Effect of Seed-Placed Ammonium Sulfate and Monoammonium Phosphate on Germination, Emergence and
Early Plant Biomass Production of Brassicae Oilseed Crops

165

Canola Council of Canada. 2010. Sulphur in Soil fertility and canola nutrition, Soil fertility
(chapter 9), Canola Growers Manual. Winnipeg, MB Canada R3B 0T6.
http://www.canola-council.org/contents9.aspx

Follett, R. H., Murphy, L. S., and Donahue, R. L. 1981. Fertilizers and soil amendments.
Prentice-Hall Inc., Englewood Cliffs, NJ. 557 pp.

Grant, C.A., Rakow, G., and Relf-Eckstein, J. 2008. Impact of Traditional and Enhanced
Efficiency Phosphorus Fertilizers on Canola Emergence, Yield, Maturity, and
Quality in Manitoba. International Plant Nutrition Institute (IPNI), Brazil
http://www.ipni.net/far/farguide.nsf/$webindex/article=87B251F4062575700078
E81522B647CF!opendocument

Hall, B. 2007. Starter Fertilizers with Canola – Too Much of a Good Thing? Ontario Ministry
of Agriculture, Food, and Rural Affairs
http://www.omafra.gov.on.ca/english/crops/field/news/croptalk/2007/ct-
0307a10.htm

Harapiak, J. 2006. Maximizing seed and seed-row fertilizer benefits. Top Crop Manager (online paper)
http://www.topcropmanager.com/index.php?option=com_content&task_content
&task=view&id=896&Itemid=182

Jackson, G.D. 2000. Effects of N and S on canola yield and nutrient uptake. Agron. J. 92: 644-
649.

Janzen, H. H., and Bettany, J. R. 1984. Sulfur nutrition of rapeseed: I. Influence of fertilizer
nitrogen and sulfur rates. Soil Sci. Soc. Am. J. 48:100-107.

Karamanos, R. E., Goh, T. B., and Poisson, D. P. 2005. Nitrogen, phosphorous and sulfur
fertility of hybrid canola. J. Plant Nutr. 28: 1145-1161.

Keeney, D.R. and D.W. Nelson. 1982. Nitrogen-inorganic forms . In A.L. page, R.H. Miller
and D.R. Keeney (Eds). Methods of soil analysis: part 2, chemical and
microbiological properties. Agronomy Mono. 9, A.S.A. Madison, W I., USA pp.
643-698

Lauzon, J. D. and Miller, M. H. 1997. Comparative response of corn and soybean to seed-
placed phosphorus over a range of soil test phosphorus. Commun. Soil Sci. Plant
Anal. 28 (3/5): 205-215.

Malhi, S.S., Gan, Y., and Rancey, J.P. 2007. Yield, Seed Quality, and Sulfur Uptake of
Brassica Oilseed crops in Response to Sulfur Fertilization. Agron. J. 99:570-577.

Miller, M.H., Bates, T. E., Singh, D., and Baweja, A. J.. 1971. Response of corn to small
amounts of fertilizer placed with seed: I. Greenhouse studies. Agron. J. 63:365-
368.

Qian, P and Schoenau, J. J.. 2007. Using anion exchange membrane to predict soil available
N and S supplies and impact of N and S fertilization on canola and wheat growth
under controlled environment conditions. Pedosphere 17: 77-83.

Qian, P., and Schoenau, J. J. 2010. Effects of Conventional and Controlled Release
Phosphorus Fertilizer on Crop Emergence and Growth Response Under Controlled
Environment Conditions. Journal of Plant Nutrition, 33: 1253-1263.

Qian, P., Schoenau, J. J., and Karamanos, R. E. 1994. Simultaneous extraction of available
phosphorus and potassium with a new soil test: a modification of Kelowna
extraction. Commun. Soil Sci. Plant. Anal. 25 (5&6): 627-635

Qian, P., Schoenau, J. J., and Huang, W. Z. 1992. Use of ion exchange membranes for the
    routine soil testing. Comm. Soil Sci Plant. Anal. 23(15 & 16): 1791-1804
SAS Institute. 2008. Version 9.2. SAS Inst. Inc., Cary, NC.

7

Adaptability and Sustainable Management of High-Erucic *Brassicaceae* in Mediterranean Environment

Federica Zanetti[1], Giuliano Mosca[1], Enrico Rampin[1] and Teofilo Vamerali[2]
[1]Department of Environmental Agronomy and Crop Sciences,
University of Padova, Padova,
[2]Department of Environmental Sciences, University of Parma, Parma,
Italy

1. Introduction

The use of high-erucic acid oils is currently receiving increasing attention, due to the great interest in chemical compounds derived from "green feedstock". At world level, the production of these raw materials is constantly growing, and a real niche market has progressively been created (Mosca & Boatto, 1994). This scenario will allow greater substitution of chemicals with "green" compounds, and the introduction of industrial oil-crops may lead to further expansion of the green market. Alternative uses of crops for non-food purposes may be an interesting source of profit for farmers, as is happening for high-erucic acid oils. The current demand for these oils is still limited: at world level, it is nearly 20,000 tonnes of erucic acid, corresponding to about 57,000 tonnes of oils, used for deriving erucamide and various others chemical compounds (Figure 1).

Erucic acid is an unsaturated fatty acid (C22:1) with a large number of applications in the chemical industry because it confers desirable technological characteristics, such as high lubricity, cold stability and fire resistance, on oils and derived compounds.

Erucic acid is mainly transformed into erucamide, a slip agent for plastic film production. There are also other important derived compounds, such as behenic, brassilic and pelargonic acids, obtained through chemical reactions (e.g., hydrogenation, ozonolysis), which have interesting and innovative uses in the manufacture of chemicals, lubricants and detergents (Cardone et al., 2003; Gunstone & Hamilton, 2001; Taylor et al., 2010).

The actual world need for seeds containing erucic acid is not very large, about 100,000-120,000 tonnes, but the positive trend observed in the last few years should allow significant extension of these cultivations. This implies that studies on the adaptation of species containing erucic acid in various environments are really essential.

Brassicaceae are the most interesting botanical Family for producing erucic acid, due to the large number of suitable species and varieties, providing on their own the whole amount of erucic needed worldwide. The content of erucic acid ranges greatly, with high inter- and intra-specific variations (Table 1): within the same species, variations may be very large,

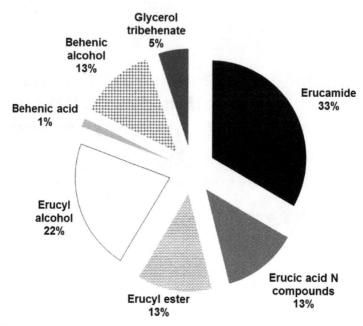

Fig. 1. World production of chemical compounds derived from erucic acid, expressed as percentages of total amount (Source: Gunstone & Hamilton, 2001).

Species	Oil (%)	Erucic acid (%)
Brassica carinata A. Braun	37–51	35–48
Brassica juncea (L.) Czern	35–45	18–49
Brassica napus L. var. *oleifera* Metzg (HEAR)	35–45	45–54
Crambe abyssinica Hochst ex R.E. Fries	30–35	55–60
Eruca sativa Mill.	28–30	34–47
Sinapis alba L.	25–30	33–51

Table 1. Variations in oil and erucic acid contents in various *Brassicaceae* species (Source: Mosca, 1998).

even higher than 30%, as for *Brassica juncea*. This probably indicates that efficiency in erucic acid accumulation may be greatly improved by adequate screening of genotypes, but also that important genetic resources may be used for breeding purposes. In this regard, an amount of 66% of erucic acid in oil must be considered, at least in rapeseed, as the current theoretical limit of accumulation (Renard et al., 1994).

For some of these species (e.g., *Brassica napus* var. *oleifera*, *Crambe abyssinica*) yield potential and environmental adaptation have been sufficiently studied in southern Europe (Lazzeri et al., 2009; Zanetti et al., 2006 a), although there is lack of information on some others, such as *Brassica juncea* and *B. carinata*.

1.1 Species of interest

In this review, four *Brassicaceae* species are considered as the most promising for producing erucic acid, i.e., *Brassica napus* var. *oleifera*, *Crambe abyssinica*, *Brassica juncea* and *B. carinata*. Although *B. napus* var. *oleifera* and *Crambe abyssinica* are the most frequently cultivated crops for producing erucic acid, because of their high and stable yields, the introduction of new species may be an opportunity to satisfy further future needs and perhaps to extend the cultivation basin of erucic acid.

The genotypes of winter *Brassica napus* var. *oleifera* are commonly defined HEAR (High Erucic Acid Rapeseed). HEAR – among the sources of erucic acid – is the most widely cultivated at world level (~30,000 ha in the world, 20,000 of which only in the UK) (Meakin, 2007). HEAR is characterised by an elevated content of erucic acid in oil, commonly over 48% (Meakin, 2007). This species can constantly provide high seed yields (De Mastro & Bona, 1998) – about 3.5-4.5 t DM ha^{-1} with 35-45% of oil – although new recently released hybrids may even reach 50% of oil (Zanetti et al., 2009). The most valuable characteristic of winter HEAR is high tolerance to low temperatures, down to –13°C (Auld et al., 1984), thus allowing early autumn sowing, long crop cycle and potentially high seed yields, even in northern Italy. With respect to the other species considered in this review, HEAR has undergone massive breeding, and the market has now made available not only open pollinated varieties but also some new CHH (Composite Hybrid Hybrid) hybrids which may reach greater yields, especially under high input management (Zanetti et al., 2006 b).

Crambe abyssinica (crambe) is one of the species richest in erucic acid (~55% of oil), but its limited cold resistance only allows almost exclusively spring sowing in north Italy and higher latitudes, with lower yield potentials (Zanetti et al., 2008). This crop may be an interesting source of erucic acid only in those environments where autumn or early spring sowings are possible. In locations with mild winters, some drawbacks may also be encountered, a marked reduction in yield being possible in cases of sudden cold spells late in the season.

Compared with the other considered species, crambe has some morpho-physiological peculiarities, such as seeds enclosed in hulls (Figure 2) – they normally remain on the seed after harvest - smaller plants, and lower oil content (~35%).

Compared with the above-cited high-erucic species, *Brassica juncea* (Indian mustard) owns peculiar morphological traits, such as taller plants, smaller seeds and elongated leaves (Figure 3). Its cultivation is widespread, especially in its native geographic area (India and Pakistan), due to some interesting agronomic characteristics (e.g., drought resistance) and the low incidence of pod shattering, which can significantly reduce yield losses. Unluckily, its resistance to cold, which is extremely differentiated among varieties, is generally quite low, although some genotypes tested can tolerate down to –5°C and provide interesting yield results. At this time in Europe there are no available commercial varieties of Indian mustard, since its cultivation is widespread only in native regions. Great interest in this species in last few decades has emerged in Canada, due to its better adaptability to spring sowing in dry environments with respect to conventional canola (Getinet et al., 1996).

Among new *Brassicaceae* suitable for cultivation in southern Europe, *Brassica carinata* (Ethiopian mustard) (Figure 4) may be considered the most promising, because it provides good seed yield (2.5-3.6 t ha^{-1}) and has several favourable characteristics, such as low bird predation and good tolerance to pests, diseases (i.e., blackleg and *Alternaria* leaf spot)

Fig. 2. Seeds of *Crambe abyssinica*, enclosed in hulls, at seed filling stage.

Fig. 3. Experimental plot with *Brassica juncea* in spring sowing (University of Padova experimental farm).

Fig. 4. *Brassica carinata* plants at flowering, characterised by simultaneous presence of white and yellow flowers.

(Bayeh & Gebre Medhin, 1992; Gugel et al., 1990; Yitbarek, 1992) and drought (Monti et al., 2009). This species has been studied since the early 1980's, showing interesting performances in semi-arid temperate climates (Fereres et al., 1983; Hiruy et al., 1983). Nevertheless, its poor tolerance to cold should be carefully taken into account in order to choose the more appropriate sowing time, especially at northern latitudes, where the possibility of early frost but also of temperature fluctuations (below and above 0°C) can significantly compromise crop success (Lazzeri et al., 2009).

Within this framework, in order to achieve high and stable production of erucic acid, it is essential to identify the most productive genotypes, among available species, for each environment. In this report, seed and oil productions of some high-erucic *Brassicaceae* species, derived from field tests set in northern Italy, are presented in response to different agronomic input management. In particular, responses of *B. napus* HEAR, *B. carinata* and *Crambe abyssinica* are discussed in relation to their results obtained from multi-scale trials (years and locations).

2. Materials and methods

In this review, the results from three separate field trials, set in different years and locations, are reported and discussed. Accordingly, the three experiments are described separately.

2.1 HEAR adaptation to reducing agricultural inputs

The trial was carried out for 2 years during growing seasons 2006-07 and 2007-08 (autumn sowing) in Legnaro (45°21'N, 11°58'E, NE Italy) at the experimental farm of the University of Padova, following a split-plot experimental design with 3 replicates. Aiming at mimicking farm-scale cultivation as closely as possible, plot size was set at around 400 m² surface area and agricultural practices were applied by means of farm-scale technologies. Three commercial genotypes of *B. napus* HEAR – two open pollinated varieties: Maplus (NPZ Lembke, Germany) and Hearty (Monsanto, France), and one CHH (Composite Hybrid Hybrid) Marcant (NPZ Lembke, Germany) – were sown on 27 and 28 September respectively in 2006 and 2007 in a silty-loam soil (pH 8.4; 2.5% organic matter; 0.1% total N) by a cereal seeder. The seed dose was 4.5 kg ha⁻¹, roughly corresponding to a density of 100 seeds per m² for all varieties. Two contrasting levels of inputs, high and low, were tested, involving soil tillage, fertilisation and weed management (Table 2).

	High input	Low input
Soil tillage	Ploughing to 0.35 m depth + grubbing + harrowing	Disc-harrowing + rotary tillage
Weed management	Metazachlor 1,250 g ha⁻¹ (pre-emergence)	Mechanical weeding (one autumn passage)
Inter-row distance	0.30 m	0.45 m
Pre-sowing fertilisation (kg ha⁻¹)	N = 30 P_2O_5 = 90 K_2O = 90	N = 0 P_2O_5 = 50 K_2O = 50

Table 2. Characteristics of two contrasting agricultural managements for HEAR rapeseed.

For proper mechanical weed control, the inter-row distance was increased from 0.3 to 0.45 m in the low input regime. After pre-sowing fertilisation (30 kg ha⁻¹ N, for high input only), spring nitrogen supply was calculated following the method proposed by CETIOM (Centre Technique Interprofessionnel des Oléagineux Métropolitains, 1998), called "Réglette Azote". In brief, it recommends increasing doses of N for rising potential yields (i.e., revealed by higher shoot biomass at the end of winter); for the same value of expected yield, the lower the shoot biomass, the higher the amount of N. According to this approach, about at the end of January, the shoot biomass of 1 m² for each plot (2 replicates per plot) was sampled and weighed (fresh green matter). Biomass values were related to potential yields of 3 and 4 t ha⁻¹, chosen respectively for low and high input managements, to calculate the N dose. The resulting amount of N, applied as ammonium sulphate in mid-February, ranged between 0-20 kg ha⁻¹ in the first year and 0-60 kg ha⁻¹ in the second for low input, and between 35-50 kg ha⁻¹ (first year) and 80-130 kg ha⁻¹ (second year) for high input, depending on genotype (Table 3).

Genotype	Input level	2006-07 N (kg ha⁻¹)	2007-08 N (kg ha⁻¹)
Maplus	High	50	130
	Low	0	60
Hearty	High	50	110
	Low	20	40
Marcant	High	35	80
	Low	0	0

Table 3. Amount of spring nitrogen in two years of trial, calculated by following "Réglette Azote" method: experimental site was considered to have deep soil rich in organic matter.

During stem elongation and flowering stages, at about 7-day intervals, the N status of the crop was checked by determining the shoot nitrogen content (Kjeldahl method). Harvesting took place about in mid-June in 2007 (Hearty and Marcant June 5, Maplus June 11), and later in 2008 (June 19, all genotypes), by a combine harvester equipped with a wheat-cutting bar. Seed oil content was measured by the Soxtec-Tecator equipment (FOSS Analytical, Höganäs, Sweden) on 2 g of milled seeds, with diethyl ether as solvent.

The profiles of fatty acids were determined on methylated oil (preparation of methyl esters according to International Standard ISO 5509) (Bondioli & Della Bella, 2002) by gas-chromatography (GC 1000, DANI Instruments, Milan – Italy). Helium was used as carrier gas and a capillary column (SUPELCO SP^{TM}-2380, 30 m long, 0.25 mm outer diameter, 0.20 μm film) to separate fatty acids. The oven was set at an initial temperature of 180°C and maintained for 5 min; later, 240°C was reached by progressive increases (+5°C per minute) and maintained for 8 min, so that each analysis lasted 25 min. The gas chromatograph had a split/splitless (SL/IN) injector (250°C) with a split rate of 100:1 and a flame ionization detector (FID) (275°C).

2.2 *Brassica carinata* response to N fertilization in multi-locality trials set in NE-Italy

In order to test the possible introduction of *B. carinata* in NE Italy, extensive field trials were set during 2006-2009 in two different locations, both in the Veneto region (NE Italy). The two sites were Legnaro (45°21′N, 11°58′E), at the University of Padova experimental farm, and Rosolina (45°04′N, 12°14′E), at the experimental research center of the Veneto region called "Pò di Tramontana". The sites were only 35 km apart, but they were characterised by different soil type, climatic conditions, and agronomic management (Tab. 4). The N fertilisation strategy was differently applied in the two sites; in the soil at Rosolina,

Localities	**LEGNARO**	**ROSOLINA**
Years	2006-2008	2007-2009
Soil type	Silt-loam	Sandy-loam
Mean annual temperature	10.2°C	14.2°C
Mean annual precipitation	820 mm	639 mm
Soil tillage	Subsoiling or cultivator + disk harrowing + rotary harrowing	Spading + rotary harrowing
Pre-sowing fertilisation (kg ha⁻¹)	30-90-90 N-P₂O₅-K₂O	30-90-90 N-P₂O₅-K₂O
Sowing date	End of September	End of September
Inter-row distance	0.45 m	0.45 m
Weed management	Mechanical	Chemical
Sowing density	4.5 kg	4.5 kg
Spring N fertilisation (kg ha⁻¹)	3 doses: 0 N 100N *Reglette dose*	2 doses: *Low*: 70÷100 N *High*: 100÷140 N
Harvesting date	Mid-June	Mid-June

Table 4. Most important protocol details of experiments in Legnaro and Rosolina with *Brassica carinata*.

characterised by a medium content of OM (1.7%), three fixed not limiting N (70, 100, 140 kg ha^{-1} N) doses were compared. In Legnaro, the three N doses tested were: i) unfertilised control (0N), ii) a high dose (100 N), and iii) an intermediate dose (Reglette).

The aim of the trials was to assess the adaptability of Brassica carinata to extensive cultivation in NE Italy. For reliable results, the trials were set on a field scale (~1 ha, each plot) adopting farm-scale technologies. The experiment lasted two years (2006-08 in Legnaro, 2007-09 in Rosolina) and the varieties BRK1 (Eurogen, Italy) and ISCI 7 (Triumph, Italy) were cultivated in Legnaro and Rosolina, respectively.

After harvest, seed yield, oil and protein were revealed and compared between cultivars and localities.

2.3 Introduction of newly released genotypes of Crambe abyssinica in southern European environment

The aim of the study was to test the possibility of introducing Crambe abyssinica as a new spring crop for large-scale erucic acid production in north Italy. The experiment was carried out at Legnaro, following a completely randomised block design, with 3 replicates. For a 2-year period (2006 and 2007) three commercial varieties of Crambe, i.e., Mario (Triumph, Italy), and Nebula and Galactica (Springdale Farm, UK) were compared in large plots (~1,000 m^2). In both years, sowing took place in early spring (March 30 2006, March 15 2007) and harvesting at the beginning of summer (July 5 2006, June 30 2007). Only limited amounts of fertilisers were applied just before sowing (30, 90 and 90 kg ha^{-1} of N, P$_2$O$_5$ and K$_2$O, respectively). Inter-row distance was 0.23 m and the amount of seeds 18 kg ha^{-1}. This high seed dose was necessary to compensate the lower germination rate of available seeds (Lazzeri, 1998). There was no need to control weeds or pests during crop cycle.

In 2008, when Galactica and Nebula were no longer available on the market, an experiment was set up to assess the most suitable sowing date. The cv. Mario was cultivated in big strips (~3,500 m^2) at two sowing times: late March (March 20) (suggested as optimal) and about the end of April (April 28) (late sowing), adopting the same seed density (15 kg ha^{-1}), and tillage system and fertilisation (30, 90 and 90 kg ha^{-1} of N, P$_2$O$_5$ and K$_2$O, respectively). Harvest was on July 3 for optimal sowing time and three weeks later (July 24) for delayed sowing.

In both experiments seed yield, oil and seed weight were evaluated at the end of the crop cycle for all tested genotypes.

3. Results and discussion

3.1 HEAR adaptation to reducing agricultural inputs

HEAR genotypes of Brassica napus turned out to be suitable for agricultural input reduction. Reduced tillage may conveniently be adopted for these new cultivars, characterised by high, fast germination. In addition, the great vitality, vigour and rapid growth re-start after winter also compensate possible reduced plant density after severe frosts. These positive traits are even more evident in hybrids, which have more vigour than open pollinated varieties. The aim of this study was to compare the adaptability of different genetic types of HEAR, open pollinated varieties vs. hybrids, in response to low input management.

The weather conditions of the two experimental seasons did not greatly differ, presenting almost the same amount of precipitation (~550 mm, during crop cycle) and similar temperatures. The only appreciable difference regarded the distribution of rain during crop cycle: the first year (2006/07) was characterised by severe drought stress during the blooming period, with only 5 mm rain in April; after two days of abundant rain at the beginning of May, the rest of the cycle was characterised by the absence of precipitation, leading to anticipated maturity, nearly two weeks with respect to the common harvest time for that environment. Instead, the second year was characterised by appropriate rain distribution over crop cycle, a fact that did not lead to any final yield increase, which was unexpectedly significantly higher in the first year (P<0.05, data not shown). Anticipation of phenological phases in the first year allowed the crop to accumulate higher and earlier aboveground biomass, probably the reason for the higher yield performance.

Analysing the effects of input reduction on main productive parameters (i.e., seed yield, oil and protein content) it emerged that cost reduction may be extremely profitable, especially if the best genotype is chosen correctly. The application of low input regimes did not lead to any significant yield reduction (P<0.05) nor of oil content (Table 5), although seed proteins were reduced, probably because of lower N fertilisation. Considering seed yield, Hearty and Marcant showed the same values under high and low inputs (3.54 vs. 3.57 t DM ha^{-1}, high and low, respectively, for Hearty; 3.38 t DM ha^{-1} in both regimes for Marcant); whereas in Maplus a limited, but not significant, yield increase in response to high input was measured (3.48 vs. 3.36 t DM ha^{-1}, for high and low input, respectively).

	Yield (t DM ha^{-1})			Oil content (% DM)			Protein content (% DM)		
	H	L	Mean	H	L	Mean	H	L	Mean
Maplus	3.48	3.36	3.42	46.1 b	47.9 a	47.0	23.8 a	22.2 b	23.0 a
Hearty	3.54	3.57	3.56	47.1 ab	46.8 ab	47.0	22.2 b	22.0 b	22.1 b
Marcant	3.38	3.38	3.38	47.0 ab	47.7 ab	47.3	22.7 ab	21.9 b	22.3 ab
Mean	3.47	3.44		46.7	47.5		22.9 a	22.1 b	

Table 5. Performance of HEAR genotypes at Legnaro (2-year means), under high (H) and low (L) input managements. Letters: statistically significant differences for multiple comparisons ('genotype × input level' interaction). Bold letters: significant differences among grand means for each variable (main effects: input level and genotype) (P<0.05, Duncan's test).

As expected, seed oil content was even more stable than seed yield, in response to intensified management but our genotypes were able to reach very high values (47.1% DM, general mean) compared with older varieties. Only Maplus appeared sensitive to input intensification, reaching significantly higher seed yield when high inputs were applied (P<0.05). Instead, seed protein content benefited by intensification (Tab. 5) (high vs. low input: 22.9 vs. 22.1%, P<0.05), due to higher N availability, as widely reported in the literature for rapeseed and other crops (Zhao-Hui et al., 2008).

One of the most important qualitative traits of HEAR production – fatty acid profile of extracted oils – appeared to be mainly under genetic control rather than influenced by agronomic management. The fraction of erucic acid in oils was only slightly higher in low input conditions (49.3 vs. 48.9% DM, low vs. high input, respectively), but differences were not statically significant (Figure 5).

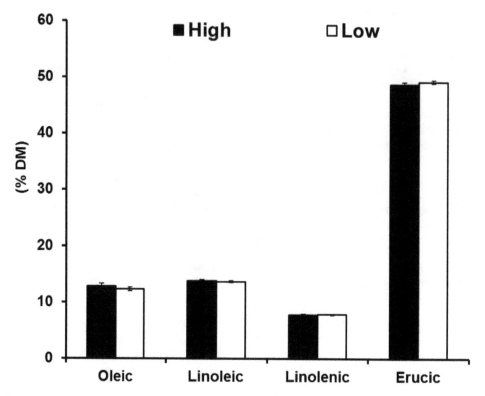

Fig. 5. Oil composition as % of main fatty acids in HEAR rapeseed cultivated under high and low input management (mean of three varieties). Vertical bars: standard error.

Analysing the oil composition of each cultivar separately (main effect genotype) (Figure 6) some significant differences emerged for oleic, linolenic and erucic acids. Anyway, the mean erucic acid content in our study was very high, always exceeding 47%. Among genotypes, Hearty was the richest and Maplus the poorest (P<0.05) (Figure 6).

It was also particularly interesting to see how Hearty oil composition was completely indifferent to input application, without any significant changes for the fatty acids in question (i.e., oleic, linoleic, linolenic and erucic acids; data not shown). In the hybrid Marcant, oil composition was found particularly interesting because of its high erucic and oleic contents and lower degree of PUFA (Poly Unsaturated Fatty Acids), normally related to higher oil stability (Zanetti et al., 2009). This peculiar characteristic of Marcant may be favourably exploited by the oil extraction industries which normally look for this kind of oil – stable and very rich in erucic acid.

We conclude that input reduction may be recommended for HEAR rapeseed without compromising yield, occasionally allowing the seed oil percentage and erucic acid accumulation to increase.

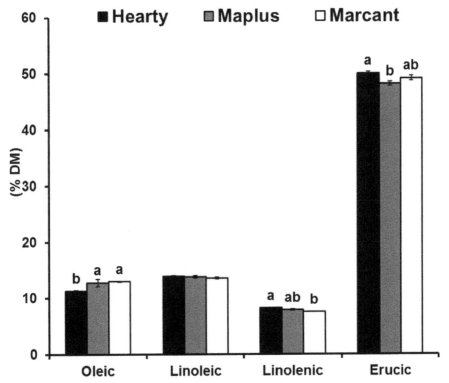

Fig. 6. Oil composition for main fatty acids in HEAR rapeseed cultivated under two input levels (main effect: cultivar). Vertical bars: standard error. Letters: statistically different values (P<0.05, Duncan's test), within same fatty acid for different genotypes.

3.2 *Brassica carinata* response to N fertilisation in multi-locality trials set in NE Italy

Brassica carinata turned out to be only partially adapted to the environmental conditions of NE Italy, since mild winters may induce early stem elongation, thus exposing the crop to severe damage during late frosts. For this reason, ascertaining the optimal sowing date for each cultivar is essential, since great variability among genotypes and localities was observed in our experiments. Our trials aimed at verifying whether the cultivation of *B. carinata* is feasible in north Italy and whether the cultivation technique used for oilseed rape is also suitable for this newly introduced oil crop.

Yield results were quite different between localities and, within the same site, between years, indicating that genetic breeding is not yet sufficient to achieve adequate stability, with the aim of providing a stable productive range for farmers.

The soil conditions at Legnaro appeared to be better for Ethiopian mustard than at Rosolina, since all production parameters considered, i.e., seed yield, oil content, and seed proteins, were significantly higher. However, the better adaptation of cv. BRK 1 (at Legnaro) than ISCI 7 to autumn sowing cannot be excluded.

The two localities differed mainly in soil type (Legnaro vs. Rosolina: silt-loam vs. sandy-loam) than in climatic conditions (Table 6. Source: ARPAV). In both localities, some climatic differences emerged between years; in Legnaro, the total amount of precipitation was similar between years, but distribution during the season was different. The first year (2006/07) was characterised by severe drought during blooming, as happened in the HEAR experiments (see above), and only 5 mm of rain fell in April; temperatures did not differ greatly between years, but both minimum and maximum temperatures during the second season (2007/08) were higher. In Rosolina, the 2007/08 season was characterised by a lack of 150 mm of water compared with both the historical mean and the Legnaro site; temperatures ranged within normal reference values for the locality (min 6.4°C; max 14.7°C). The second year in Rosolina (2008/09) was characterised by abundant precipitation, almost double that of the first season, together with higher temperatures (+1°C for both min and max).

Localities	Year	Rainfall (mm)	Mean minimum T° (°C)	Mean maximum T° (°C)
Legnaro	2006/07	588	7.2	17.2
	2007/08	548	5.7	14.8
	Mean 1992-05	593	5.4	14.9
Rosolina	2007/08	370	6.4	14.7
	2008/09	790	7.5	15.4
	Mean 1992-05	523	6.3	14.5

Table 6. Most significant climatic data from two experimental sites during two growing seasons (Oct-June) of *Brassica carinata* (Source: ARPAV)

The effect of N fertilisation was limited in this oil crop, although some influences were observed at Rosolina only. *B. carinata* seems to be a more rustic crop than *B. napus*, and a reasonable reduction of N supply would be possible to reduce cultivation costs, at least in the more fertile soil of Legnaro. In Rosolina, the effect of N fertilisation was evident in all parameters, positively influencing seed yield and protein content, but with negative effects on oil accumulation.

At Legnaro, the performance of BRK1 was quite satisfactory in terms of seed yield, oil and protein production, while in Rosolina, results fluctuated among years and N rates (Table 7). The abundant fraction of sand in the soil at Rosolina probably makes precise N management difficult to be defined.

Unexpectedly, the parameter most influenced by N fertilisation was seed oil content, which also showed an unstable response across years.

Although *B. carinata* reached higher yield at Legnaro, it was not significantly different from that at Rosolina (Legnaro and Rosolina: 2.25 vs. 1.99 t DM ha-1). Significant differences emerged for oil content, which was higher in Rosolina (P<0.05), and for seed protein

	LEGNARO					ROSOLINA					P value
	0 N	Reglette	100 N	P value	Mean	70 N	100 N	140 N	P value	Mean	Main effect site
Yield (t DM ha⁻¹)	2.45	2.53	1.78	ns	2.25	1.35 b	1.93 ab	2.71 a	*	1.99	ns
Oil (% DM)	34.7	32.1	35.7	ns	34.2 b	37.5 b	40.0 a	39.2 ab	*	38.9 a	***
Proteins (% DM)	29.7	31.8 a	32.1	ns	31.2 a	24.2 c	28.0 b	30.7 a	***	27.6 b	**
TKW (g)	4.25	4.30	3.70	ns	4.08	4.63 a	2.86 b	3.19 b	*	3.56	ns

Table 7. Seed yield, oil content, protein content, and 1,000 kernel weight (TKW) achieved by *B. carinata* in two experimental sites (2-year means). Letters: statistically significant differences among N rates within same locality (P<0.05, Duncan's test). Bold letters: significant differences between localities for each parameter.

ns: P>0.05; *; P<0.05; **; P<0.01;***; P<0.001.

content, higher in Legnaro (P<0.05) (Table 7). A similar trend was detected in winter *B. napus*, grown in the same localities and years, with again higher seed yield and protein content in Legnaro and better oil content in Rosolina. The productive performance of *B. napus* was significantly higher than that of *B. carinata*, with ~60% greater seed yield and ~18% greater oil content. As reported in the literature (Getinet et al., 1996), Ethiopian mustard seeds were found richer in protein (+ ~58%, with respect to *B. napus*).

3.3 Introduction of newly released genotypes of *Crambe abyssinica* in southern Europe

The growing season of crambe (April-June) of the two experimental years was characterised by different weather conditions. In the 2006 crop cycle, the mean minimum and maximum temperatures were 9.8 and 19.8 °C, respectively, and precipitation was 193 mm. In 2007, mean temperatures were much higher than in 2006 (minimum: 11.8 °C; maximum: 22.8 °C), and were associated with limited rainfall in April (flowering stage) and a surplus (almost double that of the historical average for the location) in May-June (capsule filling-maturation stages) (total 276 mm, during crop cycle).

Yield (capsulated seeds) was much higher in 2006 than in 2007 (mean of varieties: 3.1 vs. 2.3 t DM ha⁻¹), probably due to the more favourable climatic conditions, i.e., warmer climate and higher rainfall at flowering. No significant differences emerged among genotypes in terms of yield, although cv. Mario, which is an Italian selection, performed slightly better in both years (Figure 7).

The mean oil content of de-capsulated seeds of crambe was comparable to that of HEAR. For instance, in 2006, Galactica reached a maximum of 47% of oil, a value commonly found in high-erucic oilseed rapes in the same environment (Zanetti et al., 2009). In 2007, the high temperatures during the crop cycle very probably caused a reduction in the final oil

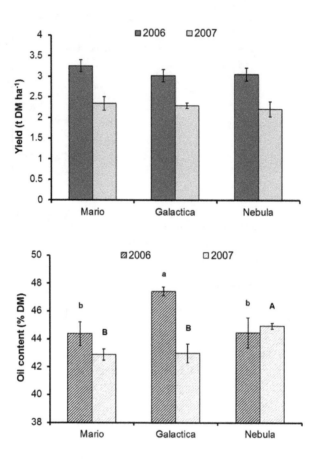

Fig. 7. Capsulated seed yield (above) in varieties of *C. abyssinica* in two-year trial and oil content (below). Letters: significant differences among varieties within same year (P≤0.05, Duncan's test). Vertical bars: standard error.

percentage of Mario (-2%) and Galactica (-5%), but not in Nebula, which showed very stable behaviour across years.

As expected, the content of erucic acid in the oil was very high in all varieties, and in 2006 reached a mean fraction of 58.4%, much higher than that of HEAR (~50%). These results indicate that, in our environment, crambe has lower productivity than HEAR in terms of erucic acid (0.61 vs. 0.89 t ha⁻¹), although the higher fraction may facilitate its separation from the other fatty acids.

In view of the short cycle of this crop in spring sowing, the total amounts of both oil and erucic acid produced seems considerable, although significant variations may be expected across years. Greater precocity may allow some genotypes to perform better in terms of oil content but not of seed yield (e.g., Nebula).

The results of the last trial (2008) on different sowing times validated this observation, indicating that the choice of a correct sowing date is essential for final results. A delayed sowing date may cause not only a decrease in final seed yield but also limited oil accumulation. The resulting oil yield may decrease by as much as ~60%, i.e., 1.12 vs. 0.46 t ha^{-1} of oil, comparing optimal and delayed sowing dates (Table 8).

Sowing date	Seed yield (t ha^{-1} DM)	Oil content (% DM)	Oil yield (t ha^{-1} DM)
Optimal (March 20)	2.65	42.4	1.12
Delayed (April 28)	1.16	39.8	0.46

Table 8. Yield performance of *C. abyssinica* cv. Mario, sown on two dates in 2008 at Legnaro experimental farm. Measures in duplicate.

One month's delay in sowing reduced crop cycle length by 18 days, which directly compromised final seed yield (-57%) and oil accumulation (-7%). These results might be even worse in the case of limited rainfall during seed filling, a frequent occurrence in early summer in this environment.

4. Conclusions

A possible increase in the erucic acid market is feasible only if large amounts of oils with high erucic acid content can be stably available on the market at reasonable prices. With this aim, studies on introducing new high-erucic species in promising environments and defining low input management are essential for the progress of this niche market.

From this study, the large-scale cultivation of high-erucic oilseed crops in the plain areas of NE Italy seems feasible with different performance among species. The productive results achieved by all studied species were in any cases encouraging. Although HEAR is the higher-yielding oilseed crop for this environment, *Brassica carinata* and *Crambe abyssinica* showed interesting prospects which should be supported by more intense breeding programmes. Several traits of these new species should be improved, especially yield stability (across years and environments). Currently, the main aspect to be investigated is their optimal sowing date, which seems to be the most important variable affecting yield. *Crambe abyssinica* appeared particularly interesting in view of its short spring cycle which may make it a good alternative to sugarbeet and soybean, two crops extensively cultivated in NE Italy.

All three of these oil crops (*B. napus* HEAR, *B. carinata* and *C. abyssinica*) turned out to be easily adaptable to input reduction, without significant changes in seed yield or quality. In particular, the positive response of these *Brassicaceae* to reduction of N fertilisation means that cultivation costs, which represent an important factor, can be reduced considerably. Other technical aspects, e.g., weed management, must be carefully investigated in future for these new crops, as no herbicides are yet registered on the market. Large inter-row distances and mechanical weeding also makes weed control easier.

Southern Europe seems a promising basin for the cultivation of high-erucic species, in view of the good soil fertility (high OM), mild winter temperatures, and the introduction of new hybrids (HEAR) which are particularly plastic and extremely high-yielding. With this assumption, stable achievement of 1 ton of erucic acid per hectare would be not only an optimistic dream but also a reliable goal.

5. Acknowledgements

The authors would like to thank the regional project "No food" of Friuli Venezia Giulia for funding research on HEAR and *Crambe abyssinica*. The Italian Ministry of Agriculture is also thanked for funding the project "Agrienegie" on *Brassica carinata*. Grateful thanks go to Dr. Franco Tosini, director of the Rosolina research centre, for kindly hosting experiments on *Brassica carinata*. Technical supervision by Adriano Massignan and revision of the English text by Gabriel Walton are also acknowledged.

6. References

Auld, D.L.; Bettis, B.L., & Dial, M.J. (1984). Planting date and cultivar effect on winter rape production. *Agronomy Journal*, Vol. 76, No. 2, (1984), pp. 197-200.

Bayeh, M. & Gebre Medhin, T. (1992). Insect pests of noug, linseed and Brassica in Ethiopia, *Proceedings of First National Oilseeds Workshop*, pp. 174-178, Addis Abeba, Ethiopia, December 3-5 1991.

Bondioli, P. & Della Bella, L. (2002). Sulla preparazione degli esteri metilici degli acidi grassi a scopo analitico. *La Rivista Italiana della Sostanze Grasse*, Vol. 79, No. 1-2, (Jan-Feb 2002), pp. 15-20.

Cardone, M.; Mazzoncini, M., Menini, S., Rocco, V., Senatore, A., Seggiani, M. & Vitolo, S. (2003). *Brassica carinata* as an alternative oil crop for the production of biodiesel in Italy: agronomic evaluation, fuel production by transesterification and characterization. *Biomass and Bioenergy*, Vol. 25, No. 6, (March 2003), pp. 623-636.

CETIOM, (1998). Nitrogen and rape in spring. *Oleoscope*, Vol. 48, (1998), pp. 9-26.

De Mastro, G. & Bona, S. (1998). Colza. In: *Oleaginose non alimentari*, G. Mosca (Ed.), 29-35, Edagricole – Ed. Agricole Calderini s.r.l., ISBN 88 206 4235 2, Bologna, Italy.

Fereres, E.; Fernandez, M., Minguez, I. & Dominguez J. (1983). Productivity of *B. juncea* and *B. carinata* in relation to rapeseed, *Proceedings of 6th International Rapeseed Congress*, pp. 293-299, Paris, France

Getinet, A.; Rakow, G. & Downey, R.K. (1996). Agronomic performance and seed quality of Ethiopian mustard in Saskatchewan. *Canadian Journal of Plant Science*, Vol. 76, No. 3, (Jul 1996), pp. 387-392.

Gugel, R. K.; Seguin-Swartz, G. & Petrie, A. (1990). Pathogenicity of three isolates of *Leptosphaeria maculans* on Brassica species and other crucifers. *Canadian Journal of Plant Pathology*, Vol. 12, (1990), pp. 75-82.

Gunstone, F. & Hamilton, R.J. (2001). *Oleochemical manufacture and application*. CRC Press, Boca Raton, Florida, USA, 325 pp.

Hiruy, B.; Riley, K.W., Nigatu, T. & Getinet, A. (1983). The response of three Brassica species to different planting dates and rates in highlands of Ethiopia. *Ethiopian Journal of Agricultural Science*, Vol. 5, (1983), pp. 22-33.

Lazzeri, L. (1998). Crambe. In: *Oleaginose non alimentari*, G. Mosca (Ed.), 95-101, Edagricole – Ed. Agricole Calderini s.r.l., ISBN 88 206 4235 2, Bologna, Italy

Lazzeri, L.; D'Avino, L., Leoni, O., Mazzoncini, M., Antichi, D., Mosca, G., Zanetti, F.; Del Gatto, A., Pieri, S., De Mastro, G., Grassano, N., Cosentino, S., Copani, V., Ledda, L., Farci, R., Bezzi, G., Lazzari, A., Dainelli, R. & Spugnoli, P. (2009). On Farm Agronomic and First Environmental Evaluation of Oil Crops for Sustainable Bioenergy Chains. *Italian Journal of Agronomy*, Vol. 4, No. 4, (July 2009), pp. 171-180.

Meakin, S. (2007). High erucic acid rape (HEAR). In: *Crops for industry: a practical guide to non-food and oilseed agriculture*, S. Meakin (Ed.), pp: 94-110, the Crowood Press Ltd, ISBN 978 1 86126 910 2, Marlborough, UK.

Monti, A.; Bezzi, G. & Venturi, G. (2009). Internal conductance under different light conditions along the plant profile of Ethiopian mustard (*Brassica carinata* A. Brown.). *Journal of Experimental Botany*, Vol. 60, No. 8, (Feb. 2009), pp. 2341-2350.

Mosca, G. & Boatto, V. (1994). Perspectives of European agriculture in new crops and processes: a challenge for the future, *Proceedings of 3rd E.S.A. Congress*, pp. 18-29, ISBN 2 9505124 1 0, Abano-Padova, Italy, September 18-22 1994.

Mosca, G. (1998). Oleaginose non alimentari. G. Mosca (Ed.), 162 pp., Ed agricole – Ed. Agricole Calderini s.r.l., ISBN 88 206 4235 2, Bologna, Italy.

Renard, M.; Pelletier, G. & Guerche, P. (1994). Les applications industrielles. *Oléagineux Corps gras Lipides*, Vol. 1, No.1, (1994), pp. 31–33.

Taylor, D.C.; Falk, K.C., Palmer, C.D., Hammerlindl, J., Babic, V., Mietkiewska, E., Jadhav, A., Marillia, E.F., Francis, T., Hoffman, T., Giblin, E.M., Katavic, V. & Keller, W.A. (2010). *Brassica carinata* - a new molecular farming platform for delivering bio-industrial oil feedstocks: case studies of genetic modifications to improve very long-chain fatty acid and oil content in seeds. *Biofuels Bioproducts and Biorefining*, Vol. 4, No. 5, (Oct. 2010), pp. 538-561.

Yitbarek, S. (1992). Pathological research on noug, linseed, gomenzer and rapeseed in Ethiopia. *Proceedings of First National Oilseeds Workshop*, pp 151-161, Addis Abeba, Ethiopia, December 3-5 1991.

Zanetti, F.; Vamerali, T., Bona, S., Mosca, G. (2006 a). Can we "cultivate" erucic acid in Southern Europe? *Italian Journal of Agronomy*, Vol. 1, No. 1, (2006), pp. 3-10.

Zanetti, F.; Vamerali, T. & Mosca, G. (2006 b). Responses of oilseed rape to decreasing agricultural inputs: hybrid vs. traditional variety, *Proceedings of IX. ESA congress*, Fragmenta Agronomica, pp. 253-254, PL ISSN 0860-4088, Warsaw Poland, September 4-7 2006.

Zanetti, F.; Vamerali, T. & Mosca, G. (2008). Performance of *Crambe abyssinica* as new crop for non-food uses in North-East Italy, *Proceedings of X ESA Congress "Multi-functional Agriculture"*, pp. 469-470, ISSN 1125-4718, Bologna, Italy, September 15-19 2008.

Zanetti, F.; Vamerali, T. & Mosca, G. (2009). Yield and oil variability in modern varieties of high-erucic winter oilseed rape (*Brassica napus* L. var. *oleifera*) and Ethiopian mustard (*Brassica carinata* A. Braun) under reduced agricultural inputs. *Industrial Crops & Products*, Vol. 30, No. 2, (September 2009), pp. 265-270.

Zhao-Hui, W.; Sheng-Xiu, L. & Sukhdev, M. (2008). Effects of fertilization and other
    agronomic measures on nutritional quality of crops. *Journal of the Science of Food and
    Agriculture*, Vol. 88, No. 7, (2008), pp. 7-23.

Permissions

The contributors of this book come from diverse backgrounds, making this book a truly international effort. This book will bring forth new frontiers with its revolutionizing research information and detailed analysis of the nascent developments around the world.

We would like to thank Dr. Uduak G. Akpan, for lending his expertise to make the book truly unique. He has played a crucial role in the development of this book. Without his invaluable contribution this book wouldn't have been possible. He has made vital efforts to compile up to date information on the varied aspects of this subject to make this book a valuable addition to the collection of many professionals and students.

This book was conceptualized with the vision of imparting up-to-date information and advanced data in this field. To ensure the same, a matchless editorial board was set up. Every individual on the board went through rigorous rounds of assessment to prove their worth. After which they invested a large part of their time researching and compiling the most relevant data for our readers. Conferences and sessions were held from time to time between the editorial board and the contributing authors to present the data in the most comprehensible form. The editorial team has worked tirelessly to provide valuable and valid information to help people across the globe.

Every chapter published in this book has been scrutinized by our experts. Their significance has been extensively debated. The topics covered herein carry significant findings which will fuel the growth of the discipline. They may even be implemented as practical applications or may be referred to as a beginning point for another development. Chapters in this book were first published by InTech; hereby published with permission under the Creative Commons Attribution License or equivalent.

The editorial board has been involved in producing this book since its inception. They have spent rigorous hours researching and exploring the diverse topics which have resulted in the successful publishing of this book. They have passed on their knowledge of decades through this book. To expedite this challenging task, the publisher supported the team at every step. A small team of assistant editors was also appointed to further simplify the editing procedure and attain best results for the readers.

Our editorial team has been hand-picked from every corner of the world. Their multi-ethnicity adds dynamic inputs to the discussions which result in innovative outcomes. These outcomes are then further discussed with the researchers and contributors who give their valuable feedback and opinion regarding the same. The feedback is then collaborated with the researches and they are edited in a comprehensive manner to aid the understanding of the subject.

Apart from the editorial board, the designing team has also invested a significant amount of their time in understanding the subject and creating the most relevant covers. They scrutinized every image to scout for the most suitable representation of the subject and create an appropriate cover for the book.

The publishing team has been involved in this book since its early stages. They were actively engaged in every process, be it collecting the data, connecting with the contributors or procuring relevant information. The team has been an ardent support to the editorial, designing and production team. Their endless efforts to recruit the best for this project, has resulted in the accomplishment of this book. They are a veteran in the field of academics and their pool of knowledge is as vast as their experience in printing. Their expertise and guidance has proved useful at every step. Their uncompromising quality standards have made this book an exceptional effort. Their encouragement from time to time has been an inspiration for everyone.

The publisher and the editorial board hope that this book will prove to be a valuable piece of knowledge for researchers, students, practitioners and scholars across the globe.

List of Contributors

Anthony O. Ananga and Violetka Tsolova
CESTA, Center for Viticulture and Small Fruit Research, Florida A&M University, Tallahassee, USA

Ernst Cebert, Suresh Kumar and Zachary Senwo
Department of Biological and Environmental Sciences, Alabama A&M University, Normal, USA

Joel W. Ochieng
Faculties of Agriculture and Veterinary Medicine, University of Nairobi, Nairobi, Kenya

Devaiah Kambiranda and Hemanth Vasanthaiah
Plant Biotechnology Laboratory, College of Agriculture, Florida A & M University, Tallahassee, USA

Koffi Konan
Department of Food and Animal Sciences, Food Biotechnology Laboratory, Alabama A&M University, Normal, USA

Felicia N. Anike
Department of Natural Resources and Environmental Design, North Carolina A&T State University Greensboro, USA

T. Y. Tunde-Akintunde and M. O. Oke
Department of Food Science and Engineering, Ladoke Akintola University of Technology, Ogbomoso, Nigeria

B. O. Akintunde
Federal College of Agriculture, I.A.R. and T., P.M.B. 5029, Ibadan, Oyo State, Nigeria

Anna Leticia M. Turtelli Pighinelli and Rossano Gambetta
Embrapa Agroenergy, Brazil

Zhen-hua Zhang and Hai-xing Song
Department of Plant Nutrition, College of Resources and Environment, Hunan Agricultural University, China
Hunan Provincial Key Laboratory of Plant Nutrition in Common University, Changsha, China
Hunan Provincial Key Laboratories of Farmland Pollution Control and Agricultural Resources Use, Changsha, China
National Engineering Laboratory of High Efficiency Utilization of Soil and Fertilizer Resources, Changsha, China

Chunyun Guan
National Center of Oilseed Crops Improvement, Hunan Branch, Changsha, China

Masumeh Ziaee
Islamic Azad University, Shoushtar Branch, Iran

P. Qian, R. Urton, J. J. Schoenau, T. King and C. Fatteicher
University of Saskatchewan, Saskatoon, SK, Canada

C. Grant
Agriculture and Agri-Food Canada, Brandon, MB, Canada

Federica Zanetti, Giuliano Mosca and Enrico Rampin
Department of Environmental Agronomy and Crop Sciences, University of Padova, Padova, Italy

Teofilo Vamerali
Department of Environmental Sciences, University of Parma, Parma, Italy

Printed in the USA
CPSIA information can be obtained
at www.ICGtesting.com
JSHW011810301024
72690JS00002B/31

9 781632 391193